Africa and International Crises

Robert W. Brown
Donald F. Heisel
Harvey Flad
Charles H. Lyons

Maxwell
School of Citizenship
and Public Affairs
Syracuse University

For two decades the Maxwell School has had as a major thrust of its teaching and research programs a concentration upon the world outside our national borders. In 1975 this commitment to foreign and comparative studies stands as a *sine qua non* of our intellectual endeavors. Increasingly the scholar and teacher is obliged to consider the influences on his work from outside our borders in order to diminish the culture-bound nature of the social sciences.

Maxwell's organization for teaching and research purposes emphasized discrete areas of the world: Soviet and East Europe, Latin America, Eastern Africa and South Asia. In addition, numerous faculty members have come here with interests in Western European countries and subjects. During the fifties and sixties, grants to subsidize a variety of such programs and interests came abundantly to Maxwell as to many other universities. That day is now gone — in fact, the scarcity of funds for supporting teaching and research about the rest of the world is a disturbing reality. As the pressures for neo-isolationism grow, as the energy shortage and related "crises" unfold, the university should feel even more sharply its extra-national role.

The Foreign and Comparative Studies Program is the present expression of Maxwell's awareness of the imperative for attention to foreign and international questions. This Program coordinates and supervises undergraduate and graduate concentrations which involve comparative and area studies. It encourages and motivates faculty cooperation in inter-regional research and teaching more effectively to meet the demands and opportunities of the present day.

The Foreign and Comparative Studies Monograph Series is a central part of this enterprise. It subsumes the series previously published by the Eastern African Studies Program and envisions an expansion of the output from the Latin American, South Asian, Soviet and East European and Western European faculties. Today, Maxwell has a large number of faculty members whose research interests include these areas. This series is a medium for their publishing manuscripts scheduled to appear in standard journals at a later date, monographs too long to appear in journals and yet not of book length, and other items. Scholars elsewhere are invited to submit their manuscripts for consideration for publication as parts of the series.

Chairman, Publications Committee — Eastern African Studies   James L. Newman

Editorial Advisory Committee — Foreign and Comparative Area Studies

    Guthrie Birkhead, Chairman   Robert Kearney
    Edwin Bock   Ronald McDonald
    Robert Gregory   Marshall H. Segall
    Robert Jensen

Publishing Desk
211 Maxwell School
Syracuse University
Syracuse, New York 13210
USA

AFRICA AND INTERNATIONAL CRISES

by

Robert W. Brown

Donald F. Heisel

Harvey Flad

Charles H. Lyons

FOREIGN AND COMPARATIVE STUDIES/EASTERN AFRICA XXII

Maxwell School of Citizenship and Public Affairs

Syracuse University

1976

Library of Congress Cataloging in Publication Data

Brown, Robert Wylie, 1922. Africa and international crises.
Foreign and Comparative Studies: Eastern Africa; 22
1. Petroleum industry and trade--Africa.
2. Africa--Population.
3. Food supply--Africa.
4. Education--Africa.
HD9577.A2A33   309.1'6'03   76-17820
ISBN 0-915984-19-9

ABOUT THE AUTHORS . . . .

Dr. Robert Wylie Brown is a member of the Geography Department of Rutgers University at Newark, New Jersey.

Dr. Donald Heisel is with the Demographic Division of the Population Council, New York City.

Dr. Harvey Flad is a member of the Geography Department at Vassar College, Poughkeepsie, New York.

Dr. Charles Lyons is an Associate Professor of History and Education at Teachers College, Columbia University, New York City.

TABLE OF CONTENTS

FOREWARD

This volume explores several problems of worldwide scope from two perspectives: how they have impacted on Africa and how Africa can contribute to their resolution. More specifically, the focus is on the petroleum squeeze, the formation of population policies, spiraling food needs and the conflict over educational priorities.

The papers in this volume are slight revisions of ones presented at a seminar held in the spring of 1975 by the Foreign and Comparative Studies Program at Syracuse University.

AFRICAN DEVELOPMENT--LOCKED ON OIL

by

Robert Wylie Brown

Without cheap energy, there would be no rich, industrialized
West. It has been some time since Europe's coal stoked the furnac-
es of the Industrial Revolution, but no one should forget the days
when the European powers enjoyed an energy-surplus position. Some
may argue that European countries poor in fossil fuels nevertheless
became rich, but these, notably the Scandinavian countries and
Switzerland, benefitted from their situation in an energy-rich region, as well as getting a considerable boost from cheap hydropower.
The United States and the USSR have, of course, fueled their
superpower status with coal, oil and natural gas. Even resource-
poor Japan initiated industrialization with its own coal and
hydropower.

As we have moved from the Age of Coal to the Age of Fluid
Hydrocarbons, imported fuels have played an increasing role in
supporting industrialization. Europe's postwar "economic miracle"
would not have been achieved without billions of dollars from the
United States and billions of barrels of cheap oil from the OPEC
nations. Japan's mushrooming emergence in the 1960's has significant roots in the Middle East; the Japanese energy base received
a 39% contribution from imported oil in 1961, rising to 71% in
1971. The Western development model, which has been peddled
worldwide, assumes the availability of cheap energy either local,
or imported.

But recently we have seen the price of oil in the international market rise fourfold. The flow of wealth over the world map
was revolutionized in a matter of days; Western economies, already
battling inflation, reeled from the impact of unprecedented pressure on their price-control efforts. The rich West and Japan,
however, have built their industrial base. Conservation will be
the watchword, and North Sea, Alaskan, Soviet and Chinese sources
will help the West muddle through an era of stagflation. The

dislocatións will be severe, but the standards will remain, by world measures, rich.

Meanwhile the OPEC countries, particularly those in the Middle East, will continue to float internationally Westward on their oil, investing in the industrial machine to their own benefit. The Arabization of the West may be hard to swallow, but there is a path of mutual satisfaction which may help the twain to meet and to cooperate. The West wants to maintain its industrial machine, while OPEC wants to feed it, consume its products, and, as quickly as possible, to import it.

More significantly, the world's poor nations, few of them endowed with sufficient oil for their own development, have suddenly found their fuel bills overwhelming their hopes for progress. For the seventy-five to eighty developing countries with energy-short economies, the implications of the new game are awesome. It is apparent that the oil-rich nations are not oriented toward the Third World from which they so recently emerged. According to estimates from the United States Treasury Department, only 10% of the $60 billion in 1974 oil surpluses went for loans to developing countries and for international bank obligations (Newsweek, 2/10/75); the other 90% was invested, in one form or another, in the industrial West and Japan. Meanwhile, the fuel bill in the Third World rose about $10 billion during the year (Subba Rao, 1975), indicating that about a $4 billion deficit exists between the petrodollar outflow and its return in the form of loans and investments. And to add to the depth of the impact, the cost of fertilizers, so often petroleum based, keeps rising, while the concurrent change in grain stocks also has raised the price of food to millions in the Third World.

The Impact In Africa

Looking at the gross figures for the continent of Africa, one might assume that the African peoples were in a favorable position to weather the crisis. Only fifteen years ago, after all, the total oil production in Africa was less than 300,000 barrels per day, whereas in 1974 it averaged almost 5 1/2 million barrels per day. With consumption on the continent at less than a million b/d, Africa now ranks second to Southwest Asia among world regions in the export of oil.

The difficulty is, of course, that only eight African nations are sharing in this bounty, and that the rest must purchase oil on the international market. The impact on the non-oil producers is, nevertheless, selective, dependent upon the rate of oil consumption and the availability of alternate sources of energy. The thrust of this paper will be to analyze the energy mix in African countries, classify them, consider the implications of recent trends, examine the longer term prospects for petroleum on the continent, and finally, to suggest a few "coping models" for the energy crisis.

Table I is extracted and calculated from United Nations statistics on energy consumption in Africa in 1961 and 1971, roughly approximating the designated "Development Decade". Happily, the United Nations converts its energy data into coal-equivalents so that inter-source comparisons can be made easily. The UN made no attempt, however, to estimate consumption of traditional fuels; the figures apply, therefore, almost exclusively to modern economies.

Table I documents Africa's dependence on oil. Of the 44 countries and colonies in the table, only seven, representing 18% of the total population of the continent, had, in 1971, economies that were less than 75% oil dependent. Furthermore, the role of imported oil increased over the decade. Another 16% of Africa's people were in seventeen countries whose economies were 99-100% based on imported petroleum in 1971, and the percentage of imported oil's contribution rose in fifteen cases while declining in eleven.

Consider the eleven countries which did succeed in reducing their reliance on foreign oil. Tunisia began its own production in the middle of the decade, quickly becoming a net exporter. Egypt and Angola succeeded in raising their production to the

Table 1. Energy Mix in African Countries, 1961 and 1971[1]

Country	Total Energy Production mil.met. tons Coal Equivalent[2]		Total Energy Consumption Coal Equivalent				Per Cent of Total Energy Consumption							
			1971		1961		(Net Oil Imports) %		Local Oil and Gas Production[3] %		Coal (Imported) %		Hydro-power (Imported) %	
	1971	1961	mil.met. tons	kg per capita	mil.met. tons	kg per capita	1971	1961	1971	1961	1971	1961	1971	1961
Northern Africa														
Algeria	53.03PG	20.95P	7.36	499	2.75	272	0	0	95.2	87.1	(4) 4.2	(8.9) 11.7	.5	1.2
Egypt	19.83P	5.10P	9.70	285	7.40	278	0	(25.3)	87.6	69.0	(5.5) 5.5	(4.1) 4.1	6.9	1.7
Libya	172.73PG	.90P	1.14	567	.38	268	0	0	100	89.9	0	(10.1)	0	0
Morocco	.76	.64	3.09	203	1.71	142	(76.7)	(67.3)	3.0	6.7	14.2	19.0	3.0	6.7
Sudan	.01	--	1.91	119	.62	50	(99.4)	(97.4)	0	0	0	(2.6)	.6	0
Tunisia	5.34P	.01	1.35	257	.73	169	0	(91.4)	92.5	1.2	(7.0) 7.0	(7.0)	.4	.3
Western Africa														
Dahomey	--	--	.10	36	.06	29	(100)	(100)	0	0	0	0	0	0
Gambia	--	--	.03	68	.01	35	(100)	(100)	0	0	0	0	0	0
Ghana	.36	--	1.65	186	.64	92	(76.1)	(90.8)	0	0	(1.9)	(9.2)	22.1	0
Guinea	.003	.001	.40	101	.30	95	(99.3)	(99.7)	0	0	0	0	.7	.3
Guinea Bissau	--	--	.06	107	.02	28	(100)	(100)	0	0	0	0	0	0
Ivory Coast	.02	.009	1.25	282	.26	79	(98.6)	(96.6)	0	0	0	0	1.4	3.4
Liberia	.03	.002	.58	371	.08	79	(95.2)	(97.5)	0	0	0	0	4.7	2.5

Table 1 - continued

Per Cent of Total Energy Consumption

Country	Total Energy Production mil. met tons Coal Equivalent[2] 1971	1961	Total Energy Consumption Coal Equivalent 1971 mil.met. tons	1971 kg. per capita	1961 mil.met. tons	1961 kg. per capita	(Net Oil) (Imports) 1971 %	(Net Oil) (Imports) 1961 %	Local Oil and Gas Production[3] 1971 %	1961 %	Coal (Imported) 1971 %	1961 %	Hydro-power (Imported) 1971 %	1961 %
Mali	--	--	.13	25	.06	14	(100)	(100)	0	0	0	0	0	0
Mauritania	--	--	.17	143	.02	22	(100)	(100)	0	0	0	0	0	0
Niger	--	--	.10	25	.03	9	(100)	(100)	0	0	0	0	0	0
Nigeria	99.83P	3.57P	3.33	59	1.68	32	0	0	87.8	66.2	5.8 (.4)	33.0	5.9	.8
Senegal	--	.003	.55	137	.33	102	(100)	(99.1)	0	.9	0	0	0	0
Sierra Leone	--	--	.27	105	.13	61	(100)	(91.8)	0	0	0	(8.2)	0	0
Togo	.001	--	.14	72	.04	30	(99.3)	(100)	0	0	0	0	.7	0
Upper Volta	--	--	.07	13	.04	10	(100)	(100)	0	0	0	0	0	0
Middle Africa														
Angola[4]	7.51	.15	.88	154	.34	70	0	(49.3)	90.2	39.1	(1.3)	(6.4)	8.6	5.5
Cameroon	.14	.12	.58	97	.28	57	(75.3)	(57.0)	0	0	0	0	24.7	42.6
Central African Republic	.005	.001	.10	60	.04	33	(93.9)	(95.2)	0	0	0	0	5.0	2.4
Chad	--	--	.10	27	.03	9	(100)	(100)	0	0	0	0	0	0
Congo	.04	.14P	.24	251	.13	164	(81.2)	0	15.8	97.7	0	0	2.9	2.3
Equatorial Guinea	--		.05	183	No figures		(100)	(100)	0	0	0	0	0	0

Table 1 - continued

Per Cent of Total Energy Consumption

Country	Total Energy Production (mil. met. tons Coal Equivalent²) 1971	1961	Total Energy Consumption Coal Equivalent — 1971 mil.met. tons	1971 kg per capita	1961 mil.met. tons	1961 kg per capita	(Net Oil) (Imports) 1971 %	1961 %	Local Oil and Gas Production³ 1971 %	1961 %	Coal (Imported) 1971 %	1961 %	Hydro-power (Imported) 1971 %	1961 %
Gabon	7.56P	1.02P	.52	1033	.07	154	0	0	100	100	0	0	0	0
Zaire	.54	.37	1.84	82	1.24	85	(54.8)	(48.5)	0	0	23.8 (17.7)	32.2 (26.3)	21.3	24.0
Eastern Africa														
Afars & Issas	--	--	.04	373	.02	259	(100)	(100)	0	0	0	0	0	0
Burundi	--	--	.04	11	.03	12	(92.3)	(100)	0	0	0	0	(7.7)	0
Ethiopia	.03	.008	1.00	40	.17	8	(95.6)	(87.3)	0	0	(1.0)	(7.5)	(3.4)	(4.6)
Malagasy	.02	.01	.50	73	.18	32	(92.2)	(91.8)	0	0	(4.6)	3.8 (2.7)	3.2	4.4
Malawi	.02	--	.22	49	.13	34	(69.3)	(48.4)	0	0	(22.7)	(50.8)	8.0	0
Mozambique	.35E	.34E	1.32	175	.94	141	(51.9)	(38.8)	0	0	46.8 (22.4)	60.3 (60.0)	1.4	.7
Rwanda	.01	.001	.04	10	.04	13	(69.2)	(97.2)	2.6		0	0	28.2	2.7
Somalia	--	--	.09	31	.04	21	(100)	(100)	0	0	0	0	0	0
Tanganyika	.04	.01	.93	70	.41	43	(95.1)	(97.3)	0	0	.3	.5	4.4	2.2
Uganda	.10E	.06E	.73	72	.21	31	(91.1)	(85.0)	0	0	0	0	8.9	15.0
Zambia	.93	.03	2.01	470	1.51	458	(33.0)	(13.2)	0	0	41.3 (.9)	71.6	25.8 (19.8)	15.2 (15.9)
Zanzibar	--	--	.02	51	.02	52	(100)	(100)	0	0	0	0	0	0

Table 1 - continued

Country	Total Energy Production mil. met. tons Coal Equivalent[2] 1971	1961	Total Energy Consumption Coal Equivalent 1971 mil.met. tons	1971 kg per capita	1961 mil.met. tons	1961 kg per capita	Per Cent of Total Energy Consumption (Net Oil) (Imports) 1971	1961	Local Oil and Gas Production[3] 1971	1961	Coal (Imported) 1971	1961	Hydro-power (Imported) 1971	1961
Southern Africa														
Rhodesia	3.79CE	3.35CE	3.38	614	2.28	606	(10.4)	(21.7)	0	0	79.3	71.9	10.3	6.3
S. Africa[5]	58.83C	39.57C	71.70	2895	43.54	2380	(20.6)	(11.6)	0	neg	79.4	88.4	.02	neg.
All Africa[4]	32.0PG	76.4P	122.1	340	70.1	248	0	0	45.3	26.8	51.8	71.0	2.7	1.7
Tropical Africa	117.6P	6.0P	22.4	90	10.7	55								
Less Nigeria	17.8P	2.4P	19.1	100	9.0	59								
Less Gabon & Angola	2.7	1.2	17.7	97	8.6	58								
Comparative Countries														
Brazil	20.4	10.2	49.5	515	24.4	339	(54.4)	(54.3)	25.2	27.1	9.5 (4.5)	9.3 (4.1)	10.9	9.7
Costa Rica	.13	.05	.8	448	.28	227	(83.8)	(81.7)	0	0	0	0	16.1	18.3
India	84.9	57.9	102.7	186	66.5	150	(16.1)	(14.0)	9.8	1.0	70.7	83.2	3.4	1.8
Phillippines	.4	.3	11.1	292	4.5	157	(96.5)	(92.9)	0	0	(.5)	(3.6)	2.9	3.5

Table 1 - continued

Per Cent of Total Energy Consumption

Country	Total Energy Production mil. met. tons Coal Equivalent[2]		Total Energy Consumption Coal Equivalent				Local Oil and Gas (Net Oil) (Imports)		Production[3]		Coal (Imported)		Hydro-power (Imported)	
	1971	1961	1971 mil.met. tons	kg per capita	1961 mil.met. tons	kg per capita	1971	1961	1971	1961	1971	1961	1971	1961
Japan	49.4	65.5	341.9	3267	122.4	1301	(.6G)(71.4)	(38.7)	1.3	1.8	(13.6) 23.4	(7.8) 52.7	3.3	6.8
Sweden	6.5	4.8	49.4	6090	26.1	3470	(81.3)(69.4)	{.8G}{.2G}	0	.5	(5.1) 3.0	(2.5)	(1.4) 13.6	17.0
USA	2029.2C	138.0C	2327.4	11,241	1478.3	8044	(14.3)(8.2)		63.8	66.5	19.4	23.8	1.7[6]	1.3[6]

[1] Source: Department of Economic and Social Affairs, Statistical Office of the United Nations, Statistical Papers Series J, No. 9 World Energy Supplies, 1961-1964 (UN,1966) and No. 16 World Energy Supplies, 1966-1971 (UN, 1973). Columns 1-6 in the table are taken directly from UN figures; other data are calculated.

[2] Net surpluses exported are indicated qualitatively with the following code: C-coal; E-electricity; G-natural gas; P-petroleum.

[3] It is assumed that locally produced oil is consumed at home, which is often not the case. But costs of imports are usually offset by equivalent quantities exported.

[4] Angola includes Cabinda, where the major oilfield is located.

[5] UN figures group South Africa with its customs union including Namibia, Lesotho, Botswana, and Swaziland.

[6] UN figures include nuclear power.

same end, leaving eight non-oil producing nations which substi-
tuted another form of energy for foreign oil. In five of these
cases (Guinea, Liberia, Togo, Central African Republic, and
Tanzania) the substitution factor was on the order of 2% or less,
which can hardly be called significant. Furthermore, in each of
these five the actual per capita energy contribution of imported
oil increased over the decade, -- for Liberia the increase was
close to 450%.

That leaves three countries, Ghana, Rhodesia and Rwanda, in
which to search for a significant reordering of the energy mix.
Rwanda's hydroelectric expansion effectively reduced consumption
of imported oil, but the UN figures show an actual decline in per
capita use of energy over the decade, putting Rwanda on the bottom
rung on Africa's energy ladder. The substitution was achieved,
therefore, at the expense of "progress" in the Western mode.
Ghana's per capita energy use doubled over the decade, with the
Volta River Project raising hydropower's contribution from 0 to
22%. In spite of this, imported oil contributed far more energy
per capita by the end of the decade than at the beginning. Hydropower's percentage contribution, in fact, declined from 30% to
22% between 1968 and 1971, even though the Volta project was
increasing its electrical output year by year over that period.

Rhodesia remains, and is clearly a special case, having been
subjected to United Nations economic sanctions since 1966 and an
Arab oil embargo (at least nominal) since 1967. While Rhodesia's
per capita energy use, which was the second highest in Africa in
1961, did not increase significantly over the decade,[1] the country
did succeed in reducing its reliance on imported oil from 21.7%
of the energy mix to a miniscule 10.4% by 1971, all of this without severe dislocations to one of the most modern economies in
Africa. Rhodesia is blessed with both cheap coal and hydroelectric

---

[1]Rhodesia was surpassed by Gabon in 1971. But the figure for
Gabon is inflated by the throughput for her regional refinery.
In ultimate consumption, therefore, Rhodesia was probably
still second in 1971.

development as alternatives, however, making this achievement hard to duplicate elsewhere on the continent.

Analysis of the data in Table I has led me to extract four major energy-related classes of countries, as shown in Table 2. Class A countries are those which are now net oil exporters, obviously poised to continue their development on a petroleum energy base with no short-term limitations. These may be subdivided into Class A_1, consisting of Libya and Gabon, which have been catapulted into the realm of rich nations on a per capita basis, and Class A_2, in which the oil revenue-to-population ratio performs no such miracles. Note that Congo has joined this group since 1971 with the exploitation of its Emeraude offshore field discovered in 1969, and Zaire should also enter this class by 1976 with production from two offshore fields now under development. Nigeria is, in fact, in a class by itself, though here assigned to the A_2's. The total capita generated by oil in Nigeria will provide opportunities for investment unavailable to the others in Class A_2, making this country more than ever the pivot of West Africa. The revenue figures estimated in the footnote of Table 2 are, incidentally, rather conservative; when the actual figures are revealed, the revenues may be closer to $10 a barrel, yielding more than $8 billion total for Nigeria.

While about a third of Africa's population is currently oil-safe or oil-rich in Class A only 9 per cent appear to be coal-safe or coal-rich in Class B. Coal has long been the pillar of the economies of Rhodesia and South Africa, contributing 79% of the energy mix to both countries in 1971. But Zambia is a new member of the coal club. As detailed by Griffiths (1968), Zambia asserted her independence from Rhodesia after 1964 by exploiting her own coal deposits, more costly than Wankie coal but nevertheless eminently usable. While imported oil in Zambia increased its contribution from 13% to 33% over the decade, coal retained predominance with 41%, less than 2% which was imported. Zambia has not been able to shake free of Kariba electric power, however; 20% of her energy was still tied to this source in 1971.

Table 2: Energy Classification of African Countries

Class A: Oil Surplus Economies

A_1 - Oil-Rich Economies ($1,000 + Per Capita in Oil
Revenues, 1974)[1]

Gabon, Libya

A_2 - Oil-Safe Economies

Algeria, Angola, Congo, Egypt, Nigeria, Tunisia

Class B: Coal-Oriented Economies

Rhodesia, South Africa, Zambia

Class C: Imported-Oil-Dependent Economies: 50-75% in 1971

Malawi, Mozambique, Rwanda, Zaire

Class D: Imported-Oil-Dependent Economies: More than 75% in 1971

D_1 - Low-Energy Economies (less than 100 kg/capita coal
equivalent from imported oil, 1971)

Burundi, Cameroon, Cent. Af. Rep., Chad,
Dahomey, Ethiopia, Gambia, Malagasy, Mali,
Niger, Somalia, Tanzania, Togo, Uganda, Upper
Volta.

D_2 - Relatively High-Energy Economies (More than 100 kg/
capita coal equivalent from imported oil, 1971)

Afars & Issas, Equatorial Guinea, Ghana, Guinea,
Guinea Bissau, Ivory Coast, Kenya, Liberia,
Mauritania, Morocco, Senegal, Sierra Leone,
Sudan

---

[1]Revenues for 1974 estimated on the basis of $8 per barrel
net exports, using production figures thru October pro-rated for
the year. Results, in round numbers: Libya, $3600; Gabon, $1100;
Algeria, $180; Nigeria, $100; Congo, $100; Angola, $75; Tunisia,
$30; Egypt, $10.

The Class C countries representing another 9% of Africa's population are transitional, with a 50% to 75% dependence on imported oil. Zaire is particularly transitional, and is headed toward Class A. With the development of hydroelectric power and a small but growing coal production, Zaire will soon join Nigeria and Algeria in that very select group of African countries which have three prongs to their energy base.[2] Mozambique is politically transitional--will independence from Portugal permit an energy exchange with South Africa, electricity for coal? Having 22% of her energy base contributed by South African coal and 80% of Cabora Bassa's initial output contracted to South Africa, independent Mozambique has inherited a strong energy link potentially offering mutual satisfaction. Newly independent nations do not always play economically rational games, however; Mozambique is faced with an interesting challenge. It is also worth noting that 1971 was the first year in which Mozambique was more than 50 per cent reliant on imported oil. Given the changing cost ration of fuels and the coming political pressures, Mozambique might follow the Zambian model and join Class B.

The case of Rwanda already has been mentioned; the only independent country whose energy consumption actually dropped, in aggregate and per capita, over the decade. Malawi, landlocked and energy poor, has shifted its orientation from imported coal to imported oil and was clearly headed for Class D in 1971. Political relationships with South Africa may determine whether a reverse shift takes place.

Twenty-eight countries are in Class D, 75% or more reliant on imported petroleum. All of them are in Tropical Africa and represent 57% of the population of that region; most are small in population or landlocked, or both. The D_1's are low-energy

---

[2]Counting petroleum and natural gas as one prong, and ignoring Morocco's tiny oil output, which has been dying, and her natural gas output, insignificant in a country of this size.

economies with weakly developed modern sectors; few of them experienced major growth in energy consumption over the decade, and
their subsistence orientation will protect most of their people
from serious dislocations due to high prices of oil. All of them
are 90% or more dependent on imported oil in their energy base,
however; Western style development in the coming decade will be
difficult in the extreme, unless alternate sources of energy are
found. The D_2's are more energy intensive than the D_1's and
subject to more immediate impact from the high cost of fuel. Three
of them experienced spectacular growth in energy consumption over
the decade--Mauritania 650%, Liberia 470%, and Ivory Coast 350%--
all of whose fuel imports were comfortably covered by expanding
exports. The 1974 balance of payments figures may tell another
story, however, since each of the three of them is more than 95%
reliant on imported oil.

Class B, C and D countries are ranked in Table 3 according to
their per capita energy consumption of imported oil in 1971.
South Africa which fuels 42% of the motor vehicles on the continent
is way out in front, in spite of the low percentage of its energy
base contributed by imported oil. Only one other B country, Zambia,
makes it into the top 15, which is otherwise populated with D_2's.
The remaining twenty are C's and D_1's, with the single anomalous
intrusion of Rhodesia, which, amazingly, slips into a slot twelfth
from the bottom, even though in 1971 it supported more motor
vehicles than Nigeria or Egypt.

The role of imported oil in Africa is quite well documented.
The North, with the exception of Morocco and Sudan, is Class A Oil
Safe, net exports providing, at least currently, a comfortable cu-
shion. White-dominated southern Africa is coal-oriented, though
South Africa needs large quantities of imported oil for gasoline
to run its relatively large vehicle population. But the broad belt
of tropical Africa has only eight countries which are not 75% or
more locked on imported oil. And while Western Africa has its
Nigeria and Central Africa its Gabon, Angola and Congo, in Eastern
Africa there is no country which is a net exporter of energy, no

Table 3:   Countries of Africa Ranked According to Per Capita
Energy Contribution of Imported Oil, 1971[1]

Rank	Country	Energy Class	% Energy Consumption in Imported Oil	KG/Cap in Coal Equivalent Generated by Imported Oil
1	S. Africa	B	20.6	596
2	Afars & Issas	D_2	100	373
3	Liberia	D_2	95.2	353
4	Ivory Coast	D_2	98.6	278
5	Equatorial Guinea	D_2	100	183
6	Kenya	D_2	92.1	158
7	Morocco	D_2	76.7	156
8	Zambia	B	33	155
9	Mauritania	D_2	100	143
10	Ghana	D_2	76.1	142
11	Senegal	D_2	100	137
12	Sudan	D_2	99.4	118
13	Guinea Bissau	D_2	100	107
14	Sierra Leone	D_2	100	105
15	Guinea	D_2	99.3	100
16	Mozambique	C	51.9	91
17	Cameroon	D_1	75.3	73
18	Togo	D_1	99.3	71
19	Gambia	D_1	100	68
20	Malagasy	D_1	92.2	67
21	Uganda	D_1	91.1	66
22	Tanzania	D_1	96.0	66
23	Rhodesia	B	10.4	64
24	Cent. Af. Rep.	D_1	93.9	56
25	Zaire	C	54.8	45
26	Ethiopia	D_1	95.6	38
27	Dahomey	D_1	100	36
28	Malawi	C	69.3	34
29	Somalia	D_1	100	31

Table 3 - continued Rank	Country	Energy Class	% Energy Consumption in Imported Oil	KG/Cap in Coal Equivalent Generated by Imported Oil
30	Chad	D_1	100	27
31	Mali	D_1	100	25
32	Niger	D_1	100	25
33	Upper Volta	D_1	100	13
34	Burundi	D_1	92.3	10
35	Rwanda	C	69.2	7

[1]Source: Calculated from UN Statistical Papers cited in Table 1. Class A countries are not listed, including Congo, which moved into Class A after 1971. As in Table 1, S. Africa refers to the customs union in Southern Africa.

country with even a smidgeon of petroleum production. Surely no
other region of the world with 100 million population finds itself
so handicapped in the current energy crisis.

Hydropower In Africa

Given the unsatisfactory distribution of coal and petroleum
on the African continent, the only form of modern energy available
to most countries is hydropower. While the potentials of African
hydroelectric power have often been exalted, the limitations are
also well known, including the rather newly-recognized ecological
consequences of dam building in the tropics. Some indication of
hydro's contributions are given in the paragraphs and tables above,
but a brief focus on hydropower in the decade under study will
further reveal the degree to which African countries have been
able to exploit its potentials.

Table 4 ranks the top ten countries in hydropower consumption
per capita in 1971. It is clear that Zambia's copper processing
is still the greatest electricity magnet on the continent, though
the Aswan High Dam has made Egypt number one in total consumption.
Comparison with Table 3 indicates hydropower's relatively small
role. Rhodesia is the only country which has been able to equate
its hydroelectric consumption to that of imported oil; and with
the exception of the top two consumers, who fed power-oriented
industries with electricity from Kariba Dam, the leaders in hydroelectric consumption parallel the bottom ten in imported oil
consumption. Hydropower did increase its share of energy consumption by 2% or more in eleven countries over the decade (Angola,
Central African Republic, Egypt, Ghana, Liberia, Malawi, Nigeria,
Rhodesia, Rwanda, Tanzania and Zambia), but there was a similar
decline in four countries (Cameroon, Ivory Coast, Uganda, and
Zaire). Over the continent as a whole, there was only a 1%
increase in hydro's share of the energy mix.

Until the current crisis, at least, Africa's greatest hydroelectric potential was not being exploited as fast as was imported

Table 4: Energy Contribution of Hydroelectric Power, 1971[1]

The Top Ten in Africa

Rank	Country	KG/CAP Coal Equiv. Produced by Hydroelectric Power
1	Rhodesia	128 (63 consumed at home)
2	Ghana	41
3	Zambia	28 (with imports, 113 consumed at home)
4	Cameroon	24
5	Egypt	20
6	Zaire	17
7	Liberia	17
8	Angola	13
9	Morocco	12
10	Congo	7

Comparative Cases, Worldwide

Norway	1900
Canada	933
Sweden	804
Switzerland	581
USA	191 (incl. Nuclear)
Japan	108
Costa Rica	72
Brazil	56
India	6

[1]Source: Calculated from UN Statistical Report cited in Table 1.

oil. While annual hydroelectric power production and consumption almost tripled during the decade under study, the actual increase was only 2.15 million metric tons of coal equivalent. Meanwhile annual imported oil consumption more than doubled, the actual increase being 19.93 million metric tons of coal equivalent, more than nine times the increase in hydro consumption. In the Development Decade, foreign oil's siren songs were too sweet to ignore.

The Outlook For Domestic Oil In Africa

Having established the primacy of oil in the African energy mix and the relative safety of the Class A oil producing nations, it remains to examine some long range prospects for petroleum in Africa. For how long are the Class A nations oil safe? Are there any other prospective members of the club? These questions cannot be answered definitely, but current trends do provide some indications.

Figure 1 shows average daily oil production for 1974. Note that Nigeria moved into first position, surpassing Libya for the first time since 1961, when Algeria was still supreme. Production in North Africa, which peaked in 1970, was still well above that of Tropical Africa in the first half of 1974, but the gap has been closing steadily since 1969. In the latter months of 1974 tropical Africa outproduced North Africa by a substantial margin, as Libyan production fell to just over one million barrels per day. While total African output in December slipped to 4.5 million barrels per day, Africa still outranked Latin America.

African reserves are also substantial. World Oil (August 15, 1974) estimates 57.2 billion barrels in proved reserves on the continent at the end of 1973,[3] which represents 10.5% of the world's total. A decade ago World Oil credited Africa with only

---

[3]Oil and Gas Journal for December 30, 1974 assigns Africa 68.3 billion barrels, with more generous figures for every country but Algeria, which is given only 7.7 billion barrels. The Journal credits Congo with a major reserve of 4.9 billion.

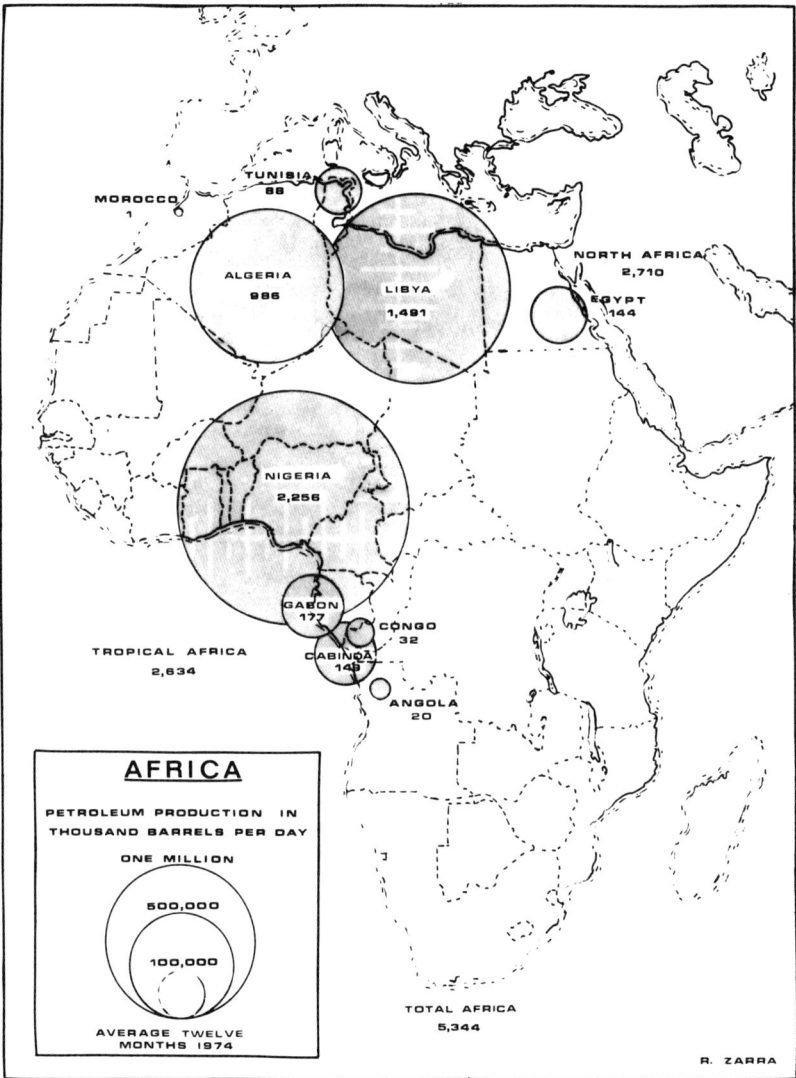

MOROCCO
1

TUNISIA
88

ALGERIA
986

LIBYA
1,491

NORTH AFRICA
2,710

EGYPT
144

NIGERIA
2,256

GABON
177

CONGO
32

CABINDA
149

ANGOLA
20

TROPICAL AFRICA
2,634

AFRICA

PETROLEUM PRODUCTION IN
THOUSAND BARRELS PER DAY

ONE MILLION

500,000

100,000

AVERAGE TWELVE
MONTHS 1974

TOTAL AFRICA
5,344

R. ZARRA

17 billion barrels, less than 5% of the world total. As Figure 2 indicates, the bulk of the reserves--more than 60 per cent--are in the north, and Libya holds the largest single share. Tropical Africa's gains in the last decade have been spectacular, however; at the end of 1963 only about a billion barrels were recognized south of the Tropic of Cancer, as compared to 21.6 billion barrels a decade later. The large circle in Nigeria represents an upgrading of reserves by almost 6 billion barrels in calendar year 1973. The map fails to show reserves in Angola and Zaire; the former are insignificant compared to Cabinda, and the latter were announced late in 1974 as being about 500 million barrels.

In view of the enormity of the total known reserves and the rapidity of petroleum development in Africa, one might assume that the continent would have a major share of the world's giant fields, those billion-barrels-plus reservoirs which are the world's most prized energy sources. Unfortunately, this is not the case. Of the 14 super-giant fields(10 billion barrels or more), none is in Africa; ten are in the Persian Gulf states, two in the Soviet Union, one in Venezuela and one in Alaska. Recent news from Mexico suggests that a fifteenth supergiant may be emerging there, the first to be discovered since Prudhoe Bay in 1968.

Of the total of 105 giants and supergiants, fully half are in Southwest Asia; USSR and North America have twenty each. Africa has thirteen giant fields. These thirteen fields contain more oil than yet remains in South America but much less than can be found in Kuwait's Burgan field or Saudi Arabia's Ghawar. By the end of 1973, almost a quarter of the recoverable reserves in these fields had been produced.

The distribution of Africa's giant fields is very north-ended. Nine are in Libya, two in Algeria, and one in Egypt; only one field Cabinda's Malongo, is located South of the Tropic of Cancer, and it barely qualifies as a giant. The absence of Nigeria as a site of giant fields is striking. The oil reservoirs in this country have been found in thin sandstone beds, only four of which were rated at more than one half billion barrels by the end of 1973. It appears

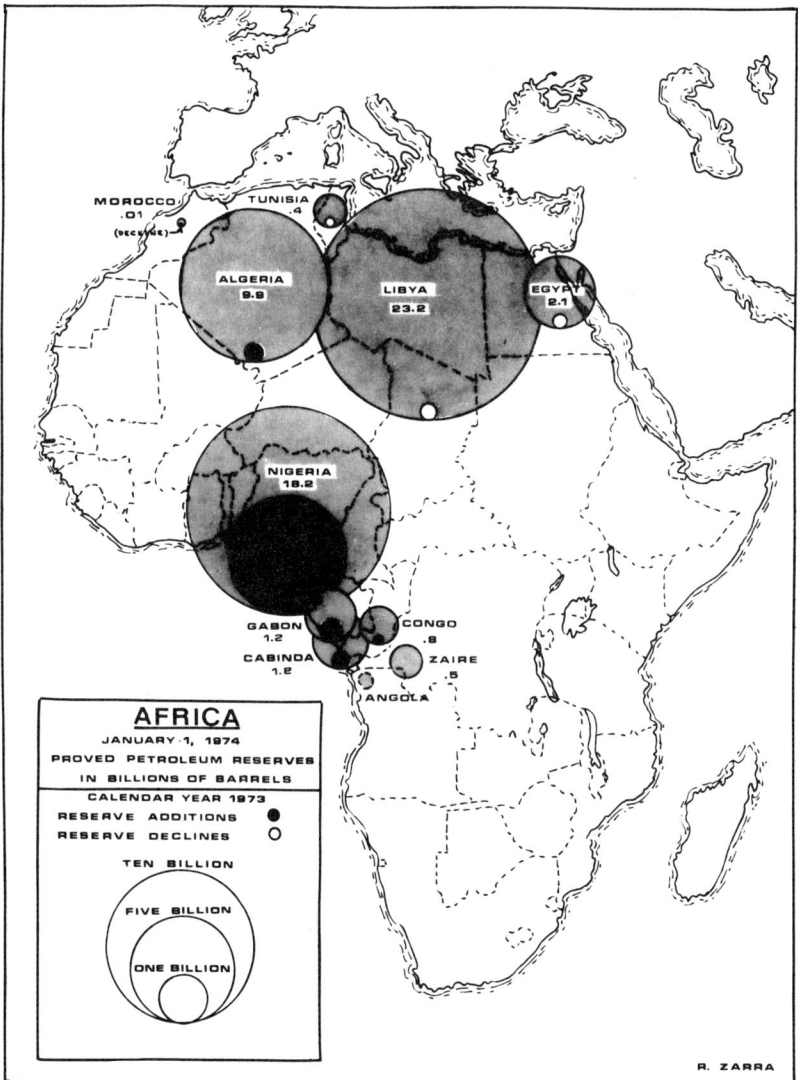

MOROCCO
.01
(DECLINE)

TUNISIA
.4

ALGERIA
9.9

LIBYA
23.2

EGYPT
2.1

NIGERIA
18.2

GABON
1.2

CONGO
.8

CABINDA
1.2

ZAIRE
.5

ANGOLA

AFRICA
JANUARY 1, 1974
PROVED PETROLEUM RESERVES
IN BILLIONS OF BARRELS

CALENDAR YEAR 1973	
RESERVE ADDITIONS	●
RESERVE DECLINES	○

TEN BILLION

FIVE BILLION

ONE BILLION

R. ZARRA

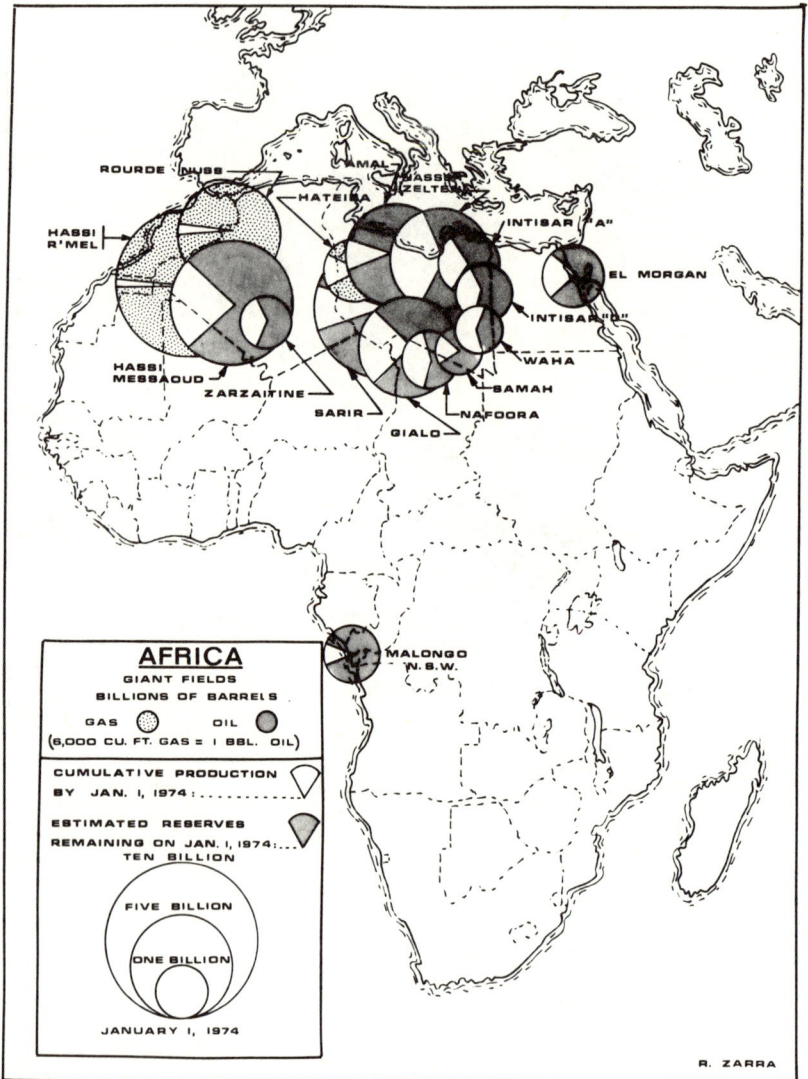

AFRICA
GIANT FIELDS
BILLIONS OF BARRELS
GAS   OIL
(6,000 CU. FT. GAS = 1 BBL. OIL)

CUMULATIVE PRODUCTION
BY JAN. 1, 1974:...........

ESTIMATED RESERVES
REMAINING ON JAN. 1, 1974:...
TEN BILLION

FIVE BILLION

ONE BILLION

JANUARY 1, 1974

R. ZARRA

unlikely that any of Nigeria's known fields will be upgraded into giants.

The map also shows Africa's three giant gas fields, drawn to energy equivalent with the oilfields. It is apparent that the two massive Algerian fields, with the equivalent of 14 billion barrels of oil between them, constitute an extraordinary source of energy which has barely been tapped, evan though exports have been occur- ing for more than a decade.

Table 5 indicates that all of Africa's known giant fields have been discovered in the last two decades; four were tapped in the 1950's and nine in the sixties. No giant discovery has been made in the last seven years,[4] however, and no field of more than two billion barrels has been found since 1961. This is true in spite of the fact that the peak drilling years in Africa were 1969 and 1970. Since then exploration activity has declined markedly in the north, particularly in Libya, while Tropical Africa, partic- ularly Nigeria, has been drilled more intensively then ever before, though a decline occurred in 1973.

The "reserve life index" of the table also shows that several of the giants are being rapidly depleted. The index is merely the result of dividing 1973 production into the remaining reserves; it has no value for prediction of field life because it does not take into account changes in production rates due to reservoir charac- teristics or to political factors, nor does it incorporate addi- tions to the reserves which may come with further development of some of the newer fields. But it does indicate relative longevity of the field's based on today's knowledge. Note that only two fields have life indexes of more than fifty years and that five are indexed at twenty-five years or less, including Tropical Africa's only giant, Malongo. The probable short life of Nigeria's two lar- gest fields is indicated in note three; none of the five biggest reservoirs in that country could sustain current rates of produc- tion for more than a quarter of a century.

---

[4]Congo's Emeraude field is known to be a giant reservoir, but at present only 6%-7% of the oil is considered recoverable.

Table 5: THE GIANT OILFIELDS OF AFRICA[1]

Reserves rank	Name of Field	Country	Year of Discovery	RESERVES Est. Proved 1/1/74[2] bil bbls	RESERVES % Exhausted	PRODUCTION 1973 mil bbls	PRODUCTION 1973 Rank in Africa[3]	PRODUCTION b/d per well 1973[4]	PRODUCTION Cumulative to 1/1/74 mil bbls	Reserve Life Index, yrs.[5]
1	Sarir	Libya	1961	8.3	9.0	83	3	4,500	752	91
2	Hassi Messaoud[6]	Algeria	1956	6.4[6]	23.5	155	1	2,400	1,545	31
3	Amal	Libya	1959	4.2	10.7	44	14	1,300	453	86
4	Zelten	Libya	1959	4.2	39.7	78	4	1,700	1,657	32
5	Gialo	Libya	1961	4.1	25.2	121	2	2,000	1,062	25
6	Intisar "A"	Libya	1967	1.6	29.0	26		4,500	474	45
7	El Morgan	Egypt	1965	1.6	24.1	44	13	2,300	385	28
8	Nafoora	Libya	1965	1.4	44.8	70	5	2,200	650	11
9	Intisar "D"	Libya	1967	1.3	35.8	60	7	11,700	462	14
10	Malongo, N.S.W.	Cabinda	1966	1.2	14.6	51	10	1,300	182	21
11	Zarazitine	Algeria	1958	1.0	42.0	15		3,700	442	41

Table 5 - continued

Reserves Rank	Name of Field	Country	Year of Discovery	RESERVES Est. Proved 1/1/74[2] bil bbls	RESERVES % Exhausted	PRODUCTION 1973 mil bbls	PRODUCTION 1973 Rank in Africa[3]	PRODUCTION b/d per well 1973[4]	PRODUCTION Cumulative to 1/1/74 mil bbls	Reserve Life Index, yrs.[5]
12	Waha	Libya	1960	1.0	57.0	53	8	2,800	585	8
13	Samah	Libya	1961	1.0	22.4	18		4,500	226	44
	Total, Giant Fields			37.5	23.6	818		2,400	8,875	35
	Total, All Oil fields			72.2	20.8	2,158		1,700	15,018	26
	Total, non-Giant Oilfields			34.7	17.9	1,340		1,400	6,143	21

Giant Fields are here defined as 1 billion barrels or more. Primary source of Data: 1974 International Petroleum Encyclopedia (Tulsa, Oklahoma: The Petroleum Publishing Co., 1974), pp. 216-220.

Includes cumulative production to 1/1/74.

Fields not in the top 15 are not ranked. Several less-than-giant fields in Africa are currently out-producing some of the giants. For the record, here is an abbreviated table on the leaders amont these (probably short-lived) fields:

Name	Country	Production 1973 mil.bbls	Rank in Africa	Reserve life index, yrs.
Defa	Libya	61	6	5
Bu Attifel	Libya	52 (est.)	9	<10
Odidi	Nigeria	48 (est.)	11	<10
Jones Creek	Nigeria	47	12	10
Bahi	Libya	36	15	6

Table 5 - continued

[4]Includes only wells actually producing in 1973, not wells shut in. Source: Oil & Gas Journal, Dec. 31, 1973, pp. 122-126.

[5]Calculated by dividing 1973 production into remaining reserves. The figure is invalid for projection but provides a useful means of comparison among fields.

[6]There is disagreement in the petroleum literature as to whether Hassi Messaoud should be treated as one or two (N. & S.) fields. I have chosen to treat it as one field, combining data published in the Petroleum Encyclopedia to that end.

The reserve life index may also be applied to a country's total reserves, although the results can be even more misleading. In the case of Egypt, for example, about half the reserves are in the El Morgan field, which in 1974 was providing less than one third of the production. The reserve life index for the United States is about 12 years, and a declining production curve has been observed since 1970; but the major Alaskan reservoirs have not been brought into production, nor have they been fully explored. The figures given for African countries must be viewed as very rough indicators, and they are based on the World Oil estimates and on average production figures for 1974: Congo, 68 years; Libya, 43 years; Egypt, 40 years; Algeria, 27.5 years; Nigeria, 22 years; Angola (with Cabinda), 20 years; Tunisia, 20 years; Gabon, 18.5 years.

All the reserves in Africa, as elsewhere, are subject to revision in the light of the multifold increase in the price of oil over the last year. No doubt some known oil in Africa can be upgraded into the "proved reserves" column at today's high prices; recoverability of some fields can be improved with increased capital investment in secondary recovery methods. The Sarir Reservoir, for example, might achieve supergiant status. It is known to have a huge amount of oil in place, but there is disagreement among engineers about the recoverability of the unusually waxy oil.

On the negative side, the top of the reservoir is almost always more easily recovered than the bottom. Egypt's El Morgan was producing at better than 300,000 barrels per day in 1970 but has now fallen to less than 80,000 b/d. An elaborate water injection system will raise production somewhat by 1976, but it is clear that El Morgan's remaining reserves are going to be much harder to extract than was the first quarter of the reservoir.

The big unknown factor in the future of oil in Africa is, of course, the discoveries that will be made in the coming years. Being unqualified to evaluate the potential productivity of African geology, I can only report the indications in the literature.

Nirod (1974) and Bird (1974) summarized the year 1973, in which
exploration activity was on the decline. In the North, Egypt, with
a new discovery in the Western Desert and another in the Gulf of
Suez, appeared to be securing her oil-safe position, which would be
further buttressed with the return of the Sinai fields now under
Israeli control. Algeria and Tunisia reported no discoveries, but
potentially important finds were made in Libya. Sudan's Red Sea
Graben was attracting exploration concessions and is viewed as one
of the more promising new provinces in Africa, but Morocco was
still an oil outcast with her dwindling, tiny reserves.

The big news in Tropical Africa comes from the Niger Delta,
with 22 new discoveries, and from the Gabon-Congo-Angola offshore
basin, which is, happily, sharing its bounty among several
countries, Zaire to be the next recipient. In addition to these
areas with tested production, promise was shown with one oil well
in Cameroon and gas in the Ogaden Basin of Ethiopia, which is also
viewed as a promising area. Exploratory wells were also being
drilled for the first time in inland basins in Chad and Mauritania.

In South Africa an eight-year drilling record with 91 (offshore and/onshore) wells, one going 20,000 feet down, has so far
yielded no commercial gas or oil. While twelve wells have had
indications of hydrocarbons, the best oil well tested at 20 barrels
per day and was assigned very limited reserves (World Oil, Aug. 15,
1974).

It is tempting to conclude that most of Africa's giant fields
have been discovered and that subsequent revelations will be
relatively small compared to what we have seen in the last fifteen
years. No replicas of Libya's prolific Sirtica Basin are seen in
the African crust, and the emergence of big oil in Tropical Africa
so far has been marked by a shortage of giant fields. While
surprises undoubtedly lie ahead, it would be hazardous to assume
that only the top of the iceberg has been revealed. The question
of obtaining risk capital must also be considered. The economical-
ly-pressed West is anxious to explore close to home, and the capit-
al-rich Middle East is not interested in finding new oil elsewhere.

Whether a high rate of exploration activity can be maintained in untested provinces of Africa through the coming decade seems doubtful.

Models For Coping With The Energy Crisis

In assessing the impact of the energy crisis in Africa, it is well to remember that Africans as a whole do not consume much energy even by Third World standards. As shown on Table I, only a handful of countries in Tropical Africa consume more energy per capita than does India, and most consume far less. With the exception of Sudan, the northern countries fit into a higher range represented by Costa Rica and the Philippines. Only South Africa approaches the energy consumption figures typical in the industrialized West, and its figures are indeed impressive as compared to Brazil's or Japan's, for example.

Nevertheless, the peoples of Africa must cope with the energy crisis. Class A countries merely have to plan. The Nigerian plan was originally simple and comprehensible - let the faucets run wide open and don't worry about the reserve life index. Nigeria needed capital, not oil in the bank and the discovery rate gave an illusion of security, even though the absence of one giant field was of some concern. By the end of 1974, however, Nigerian policy had changed, with orders going to Shell-BP and Gulf for a total cutback of 250,000 barrels per day. Conservation was the announced reason for the reduction, but OPEC pressures to maintain high prices undoubtedly played a role in the decision.

Egypt would desperately like to follow the early Nigerian plan, if she could find enough oil; recent invitations to Western capital indicate that she will make another try, since an encore to Aswan is out of the question.

Playing a totally different game is Libya, which has more than enough capital for her small population. She is putting her oil in the bank where it will support her desert economy in the future. Libya's unprecedented production surge in the late 1960's

has been exceeded only by her politically-imposed production
decline in the seventies; she has now dropped to less than one
third of the 3.3 million b/d output averaged in 1970. But the
plan is very rational - the peak figure was suicidal.

It is safe to assume that most of the Class B,C, and D
countries would thrive on the problems of the A's. Coping Model I
therefore is highly appealing - Find Oil. As delineated above,
I am persuaded that few countries are likely to succeed in this
course, Sudan and Ethiopia having perhaps the best prospects.
Nevertheless, even a little oil would be a fountain of joy in
Chad or Niger or Dahomey, and one may hope that the Class A
ranks will be expanded during the decade to come.

Coping Model II, Push Coal, is also very limited in its appli-
cation. None of the Class B's -Rhodesia, South Africa and Zambia -
has demonstrated any aptitude for oil finding, and each is fortunate to be able to pursue Model II. The only other apparent
candidate for this course is Mozambique, whose coal deposits may
not be up to the challenge.

Model III, Drive on Hydropower, has many more potential candidates. With imported oil no longer cheap, hydroelectric power's
cost advantages should prove startling in some instances, and
power-hungry industries such as aluminum and copper refining could
multiply in Africa, if they are politically welcomed and if the
demand for expansion returns. As Hance (1974) recently pointed
out, Zaire's Inga II downstream scheme from Kinshasa, which will
send electricity 1,131 miles across the country for the copper
industry, sets an interesting precedent in hydroelectric
development in Africa. Improved technology in transmission and
greater cost advantages will mean that the best sites can be
exploited to serve large areas, with regional grids a possibility,
as currently discussed in Ghana and Ivory Coast (NY Times, 1975).
The age of the Superdam should be passing, however; one of Inga's
attractions is the site's response to step-by-step development
without a giant dam. Hance also sees the possibility of substi-
tuting municipal hydroelectric plants for existing carboelectric

installations in at least ten countries (Burundi, Ethiopia, Kenya, Lesotho, Liberia, Malagasy, Morocco, Niger, Swaziland, Togo), and electrification of railways may also be expanded. While hydro-power still cannot provide the mobility furnished by petroleum, it seems likely that the next decade will be marked by a more sub-stantial substitution than occurred in the period under study for this paper. Table 4 indicates that some very attractive energy-per-capita production figures have been achieved with hydro-electric power in some European countries, though conditions there were truly extraordinary and difficult to duplicate in Africa.

Coping Model IV, Boost Export Prices, has already been tried by at least one African country in Class D. If OPEC can do it, so can I, and King Hassan of Morocco raised phosphate prices from $15 to $60 a ton in 1974, a move which, incidentally, was cheerfully imitated in Israel (NY Times, 1975). While very few countries have sufficient control of a commodity to brave Model IV alone, copper or cocoa or peanut OPEC's might be formed in Africa, perhaps extending to other regions as well. With the Western economies foundering, however, the market conditions will not be favorable for some time. Meanwhile South Africa has stumbled on her own version of a Model IV as she has reaped a few extra billions of dollars in response to the West's playing games with gold. There is a major problem with Model IV, however, even where successfully implemented - it does nothing directly to improve or cheapen the domestic energy base.

Coping Model V, Regionalize, is appealing to the Pan-African idealist. Nigeria has more than enough oil for all of Tropical Africa - can't she give her sister nations a break? As long as the West pays premium prices for her nice, clean, accessible oil, it is not likely she will, unless her sister nations can offer her something that she really needs, which will probably not occur until the Nigerian economy becomes much more sophisticated. Model V flies into conflict with nationalism and is therefore likely to remain dormant for many years to come.

Coping Model VI may prove to be the most solid course for many African nations, particularly the D_1's - Africanize. Rather than totally submitting to the Western notion of "progress," let some of the best concepts in African culture guide development. Let love of land and a well-structured society be maintained or restored as prime values; let not the motor vehicle substitute for all the friendly animals who have served so long and well. Of course you can't stop "progress," but you can stop selling it.

33

REFERENCES

Bird, Philippe, 1974. "Petroleum Developments in Central and Southern Africa." Bulletin, The American Association of Petroleum Geologists, V. 58, October, 2055-2095.

Griffiths, Ieuan L., 1968. "Zambian Coal: An Example of Strategic Resource Development," V. 58, 538-551.

International Petroleum Encyclopedia, 1974. Tulas, Oklahoma: The Petroleum Publishing Co.

Hance, W.A., 1974. "Possible Developments in Hydroelectricity in Africa in Response to High Petroleum Prices," unpublished ms.

New York Times, 1975. International Economic Survey. January 26.

Newsweek, 1975, February 10.

Nicod, Marc-André, 1974. "Petroleum Developments in North Africa in 1973." Bulletin, The American Association of Petroleum Geologists, V. 58, October, 2025-2054.

The Oil and Gas Journal, 1974. December 30, pp. 105-110, 129-148, 191.

Subba Rao, G.V., 1975. "The Predicament of Developing Countries." Saturday Review, January 25, 18-19.

AN OVERVIEW OF POPULATION IN BLACK AFRICA
by
Donald F. Heisel

This paper will present a summary of current demographic conditions and probable trends in Black Africa. The countries of
North Africa--Algeria, Egypt, Mauritania, Morocco, Sudan, Tunisia--
are omitted on the grounds that they are culturally and socially
much more closely linked to the Mediterranean and Middle Eastern
World and that their demographic experience has followed a rather
different path. Also, because of political and social differences,
the white-minority-dominated countries of Southern Africa are not
considered here. The topics to be covered include population size
and location, migration, mortality, and fertility.

Not all countries in the region have yet had a complete
census and no country has a vital registration system that covers
its total population. Most of the countries with British colonial
background have had at least one full census and some of them have
a well institutionalized decennial census program. In the countries with French or Belgian colonial experience there were some
demographic sample surveys and useful experiments with sample
vital registration schemes but the first round of complete
national censuses is just now underway. (A notable exception is
Togo, which independently carried out a census in 1970.) The
United Nations is providing the essential external support,
including both funding and advisory personnel (African Census
Program 1973). External assistance for the census program is
estimated to cost some $16,000,000.

Nevertheless, the problems impeding the census program remain
formidable. In addition to the obvious difficulties of terrain,
shortages of personnel at all levels, and lack of equipment, the
people to be enumerated often lack the very information the census
is designed to gather. Moreover, they are not always willing to
share information they do have with a census official, whose

36

motives may be mistrusted. Finally, and perhaps most important,
to the extent that demographic data have political relevance, they
become vulnerable to attempts at manipulation of results. This is
clearly demonstrated in the melancholy history of the censuses of
Nigeria taken in 1962 and 1963 (Udo 1968). Reports from Nigeria
indicate that this problem may not have been avoided in the 1973
census.

In addition to the general improvement in the supply of basic
demographic data, the capability to make use of the data in Africa
is also growing. New demographic research and training centers
have been established by the United Nations in Accra and Yaounde.
Many of the universities are expanding activities in demography;
notably included are the universities of Ghana at Legon, Ibadan
and Ife in Nigeria, the National University of Zaire in Kinshasa,
and the University of Dar es Salaam. Some of these institutions
have developed programs that lead to an advanced degree in the
subject. In addition, an African Population Association has been
founded and a scientific journal is planned to begin publication
in the near future. Several quite significant conferences
recently have been held in the region. Finally, the most important
component of the development is that the number of qualified Africans active in demographic research and teaching has increased
substantially.

Many of the results of these developments, however, will begin
to have significant impact only after a few more years have passed.
The demographic data for Black Africa still are probably of the
poorest quality for any major region of the world. The comments
which follow therefore are best estimates based on prudent analysis.

Size and Distribution

The region has a total population of approximately 250 million
people. Countries range in population size from Nigeria, which
may have as many as 80 million people, to several with less than
one million.

37

The crude density for the whole region is about 12.5 people
per square kilometer. This is low in comparison with Europe and
much of southern and eastern Asia but is slightly higher than
South America, English-French speaking North America, and the
U.S.S.R. Population is distributed very unevenly throughout the
region; moreover, the pattern of distribution is relatively complex (Hance 1970). There are zones of relatively high density
found in the region running from Senegal through Nigeria. A band
of especially high density is located along the coast inland for
about 300 kilometers from Ghana through Nigeria. A second notable
zone of higher density runs from the shores of Lake Victoria,
through Rwanda and Burundi, and into Malawi. High densities also
are found extensively on the Abyssinian Plateau, and smaller areas
occur in the Central Highlands of Kenya and around the slopes of
Mount Kilimanjaro.

Distinctively lower densities can be seen along the Horn of
Africa--Somalia, the Ogaden Region of Ethiopia, and northeastern
Kenya. Another low density area is found in Namibia and Botswana.
A third low density region extends from the coastal areas of
Gabon and the Congo on into the eastern part of Cameroon and the
Central African Republic. Finally, the Saharan Desert portions
of Chad, Niger, and Mali, are very thinly populated.

Several of the major factors influencing the distribution of
population in the region have been identified. One of the most
important is the local suitability of the climate for agriculture,
given the technologies available to the population, but this is
certainly not the only factor. The presence of diseases affecting
humans or domestic animals has in some places been a strong
barrier to settlement. Onchocerciasis and sleeping sickness zones
are typical examples. Social and historical facts   also have
been important. Persisting zones of low density in some regions
of West Africa may still be a result of the depredations of the
slave trade. In eastern Africa there are areas where control of
land by nomadic populations continues to restrict the density of

settlement. Finally, during the colonial period, the economies
were chiefly oriented toward primary production for export to the
metropolitan nation; the location of roads, railways and most
major cities reflect the persisting influence of the colonial
experience.

The most important factor leading to a change of distribution
in the modern world is the growth of urban places. In general,
the population of this part of Africa is one of the least urbanized
of the world. The proportion living in an urban place tends to
run between 10% and 20% of the total in a given country. Within
the region, eastern Africa is distinctly less urbanized than
western or middle Africa. However, urbanization is occurring at a
rapid and probably accelerating pace. According to recent estimates by the United Nations, although the region is one of the
least urbanized, the increase in the proportion urban achieves the
same high levels typical of the Third World (Goldstein 1973).
Major cities are commonly growing at 5% to 7% per year--two or
more times the national rate of population increase and some
(Kinshasa, for example) are growing at well over 10% annually.
Of course, the low base makes a comparatively high rate of growth
easier to attain; small absolute numbers added to the urban population produce large relative changes.

Migration

The growth in urban populations is a combination of their
rates of natural increase plus the results of in-migration from
the countryside; about half of all urban growth is directly
attributable to in-migration. Movement of people always has been
a characteristic of Black Africa; it is remarkable to observe the
regularity with which oral traditions narrate a history of extensive geographic movement far back into the distant past.

The modern internal migration stream in Africa mostly begins
in the villages and homesteads of the rural areas--if for no other
reason than it is in the rural areas that most of the population
yet lives. Much of the internal migration is directed toward

urban areas but movement from one rural area to another also is
very common. The patterns of migration also differ by the
intended and actual length of stay in the area of destination.
Seasonal and longer term temporary migration remain very common
in Africa.

The movement out of rural areas is not easy to explain. In
most African communities, there is a deeply felt attachment to
place of origin. Although the net flow of population is very
strongly in favor of migration away from the village and into
towns and cities, the return movement back to the villages is far
from negligible. Frequent short visits, regular contact through
traveling kinsmen and sometimes a practice of trying to arrange
so that one's children can be brought up mostly in one's natal
village, serve to maintain strong links to the rural area.

The intense links to the home village arise out of far more
than sentiment alone. Throughout most of the region, an individual's security depends heavily upon the extended kin group and
upon access to land. Both are most fully accessible in the village of origin. Indeed, under land tenure arrangements prevailing
in most countries, kinship ties and access to land are intimately
interconnected.

Nevertheless, the outflow from the villages continues. It
is prompted chiefly by the lack of economic and social opportunities for those who remain. Underemployment in rural areas is a
chronic problem nearly everywhere in Black Africa. Moreover, the
pressure to leave is sometimes more than a matter of underemployment alone. During earlier times, a variety of compulsory
arrangements were used specifically to oblige villagers to join
the migrant labor force in order to meet the manpower needs of
the colonial economy. The slave trade was, of course, the extreme
case. In modern times, natural disasters to which the local
economy was unable to respond such as the Sahalian drought have
sharply increased the temporary migration of workers out of the
affected areas. A longitudinal study of one village in southern

Niger showed that during the drought years the proportion of men
who migrated seasonally to the Coast in search of employment
increased from one-third to three-fourths (Falkingham and Thorhah
1974). In some places, political and economic conditions (for
example, land reform which leaves some landless) have permitted
virtually no alternative to departure. In general, the persistence of regional inequalities in level of development exerts continuing pressure to migrate out of the village (Amin 1974).

In a somewhat different context, the increasing proportions
who have attended schools and the spread of modernizing attitudes
make some aspects of traditional rural social structure unattrac-
tive. It is not always easy for educated youth to accept the
traditional authority of age. Finally, rural life lacks ameni-
ties such as piped water and electricity which have come to be
well known and higher desired nearly everywhere (Caldwell 1969).

As indicated, a leading destination of those who leave the
rural areas is the urban centers. The attractiveness of urban
centers to migrants, though, is not as obvious as it might seem.
The urban economy is characterized by chronic high levels of
unemployment nearly everywhere in Black Africa. (The unemployment
rate is very inadequately measured but is commonly estimated to be
roughly on the order of 20 percent.) Urban life for the poor is
far from pleasant. Nevertheless, several factors are strongly
conducive to migration into urban areas. First, although there
are high levels of unemployment, the income which can be obtained
if one is eventually successful in finding a job is most attractive in comparison with the opportunities found in near-subsis-
tence agriculture. It is often perceived as justifying time lost
in job-seeking (Harris and Todar 1970). Second, though the goods
and services obtainable in urban areas are scant in comparison
with those found in an industrialized society, they are still
greater than those in the village. In addition, many important
services are available only in urban areas: hospitals and
secondary schools are important examples. Third, in many

countries if one leaves the rural area there are few places to go other than the city. Division of the countries into ethnically homogeneous local areas where languages are not mutually understandable implies that if one leaves one's own ethnic area, there is little alternative to an urban center.

Migrants moving from rural areas to other rural areas are chiefly directed toward frontier zones, areas of resettlement, or some form of commercial agricultural or extractive development. The latter attraction plays an important but now probably declining role in the migration streams of Black Africa, but during the period of direct colonial rule, such enterprises were among the leading destinations for temporary migrants.

The prospects for planned resettlement in agriculture have attracted a considerable amount of optimistic interest. The success of some schemes such as the Gezira in the Sudan have stimulated many other attempts, not all of which have had the desired results--demographic, economic or social. The sheer volume of movement toward urban areas indicates that up to now, the attraction of new areas of rural settlement has not substantially diverted the flow of migrants toward urban areas. Indeed, in some frontier areas, there is considerable risk that land being brought into agriculture is of more and more marginal value.

Some governments have become much concerned about effects of emerging patterns of population distribution and migration in their countries. Typically, stated policy goals, where any exist, are to avoid excessive urban concentration and to achieve reasonable balance in the level of development between various districts within the country. Policies to effect these goals range from forced relocations of population--a police and bulldozers approach--to exhortation of peasants to return to the land, to elaborate schemes of alternate growth poles and rural development. In some instances, such policies have had an effect. However, the general conclusion would appear to be that the trend toward concentration of population in the largest cities continues little affected by such policies. The prospects, therefore, are that migration to cities as the chief form of internal movement will continue at a rapid and probably accelerating pace.

Contrasted to internal migration, which may well increase in intensity, international migration within the region might decline and it will almost certainly be more tightly controlled. Movement across national borders in Black Africa was comparatively free during much of the colonial period (when important economic interests were best served by encouraging an adequate labor pool for any given area and when decisions about matters such as migration were effectively made entirely outside the region). However, during recent years restriction of movements and expulsions of aliens have become much more common. In some instances, the expulsions have received international attention, especially where racial minorities have been involved. The recent experience of the Asians in Uganda is a major case in point. In other instances, however, much larger numbers of Africans from neighboring countries have been expelled. For example, in Ghana in 1969 probably over 200,000 alien Africans were abruptly ejected (Peil 1974). Relatively high levels of unemployment within a country and a political elite which is more responsive to the demands of its own citizens can make expulsion of foreigners an attractive policy. In the very short term, such expulsions inflate the number of international movers but in the medium and longer run, the level of international migration in the region appears likely to decline.

Mortality

The available evidence suggests quite clearly that mortality levels in the region remain among the highest in the world. Specifically, crude death rates are commonly estimated to be twenty per thousand or above. By way of comparison crude death rates for Asia tend to run in the region of fifteen to twenty and those of

Latin America are commonly found to be around ten. Infant
mortality rates are correspondingly high--typically estimated to
be in the area of 150 to 200 deaths in the first year of life per
1000 live births.

Mortality levels are generally slightly lower in East Africa
than in West and Central Africa. The pattern is not a strong
one, though. Ethiopia and Malawi have higher than average death
rates whereas Ghana has achieved a notably lower level of mortality. There is convincing evidence that mortality levels in urban
areas are significantly lower than in rural areas. For example,
in Ghana it was found through a careful and detailed special
survey that urban infant mortality in the late 1960's was 98 and
in rural areas was 161. In the Accra Capital District infant
mortality was estimated to be down to 56 (Gaisie 1975).

The data suggest a broad historical decline in mortality
which continues into the present. Two specific pieces of
evidence support the assertion.

First, there is indirect evidence of a decline in mortality
simply in the fact that the populations of the region are growing
very rapidly. In most countries the rate is somewhere between
2 and 3 percent per year. This is so rapid and the population of
Black Africa is of such a size that it is apparent that the rate
of increase could not have been this high for very long in the
past. For example, if the rate of increase had been as high
since 1873 as it is at present, Africa would have attained its
present size starting from a base of 20 million at that date.
All evidence indicates the population is much larger than that a
century ago. At the same time, there is no evidence to support
the proposition that there has been a rise in fertility or rates
of immigration anywhere near large enough to fully account for
the rate of population increase. Indeed, for some areas there is
fragmentary evidence indicating that during and immediately after
the period of colonial expansion in from the coast, fertility
declined slightly. A major factor thus must have been a decline
in mortality.

Supporting evidence also comes directly from demographic
analysis in those countries where data are more satisfactory. It
must be admitted that such data are not conclusive for all
countries in the region; typically, it is those countries which
have developed more rapidly and presumably have better health
conditions which also have better demographic statistics. However, reports from countries such as Tanzania and Kenya may be
taken as indicative. For mainland Tanzania, the crude death rate
was estimated to be 24-25 in 1957 and 21-23 in 1967; for Kenya
the estimates were 18 in 1962 and 17 in 1969 (Blayo and Blayo
n.d.; Henin and Egero 1972). It may also be noted that if there
were errors of estimation of mortality, they may well have produced greater underestimates in the earlier censuses.

A completely satisfactory explanation of the causes of the
mortality decline is difficult to find, especially since the
decline is not well documented. However, some leading factors may
be noted.

Programs of public health and preventive medicine have been
widespread and comparatively effective. Some diseases, such as
smallpox have been virtually eliminated while others, such as
measles (which can be lethal in Africa), have been brought under
control. However, the kinds of spectacular successes achieved in
some countries, for example with the elimination of malaria, have
not occurred in Black Africa. What seems to have happened is
that such programs have lowered the level of mortality somewhat,
and have made it possible to flatten extreme peaks of mortality
resulting from epidemic outbreaks.

It appears unlikely that the medical institutions of the
region have had any very large effect on the level of mortality
through the provision of conventional curative services. The
problem is that the medical institutions remain seriously under-
developed. An index of this is the ratio of the size of population to the supply of physicians. (The same conclusions emerge
if one uses indices based on auxiliary medical personnel.) Black

Africa is by far the most poorly equipped region of the developing
world in this regard. Countries with the best medical services
such as Kenya and Ghana, have about one physician per 10,000 population. At the other extreme, Ethiopia has a physician for each
70,000 and Upper Volta one per 90,000. In comparison, India and
Bangladesh have a physician for each 5,000, the Philippines one
per 3,000, Brazil one per 2,000 and Cuba about one per 1,000.
If the very ambitious program of expansion of medical services
described in the current Nigerian Five Year Plan is fulfilled,
the doctor-patient ratio will reach the present level of Bangladesh in 1980. But non-OPEC countries in Black Africa cannot
realistically aspire to such a level of medical services in the
near future.

Moreover, the concentration of medical personnel in towns
and cities, many in the private sector or in administrative positions, means that rural areas, where the majority of the population live, are far less well served. For example, in comparatively well-off Kenya it has been estimated that the effective
ratio is one practicing physician for each 67,000 persons in the
rural areas. Conventional curative medical services have had
some impact on mortality, but appear to have most often affected
urban areas and the more affluent members of the society.

Another set of factors which must have had a positive
influence on the reduction of mortality, but which is even more
difficult to measure, is the general socio-economic and material
progress occurring in many areas. A leading example is the
spread of pipe-borne water supplies in many urban and some rural
areas. The continuing development of an all-season road network
and air transport has meant that local crop failures or the out-
break of epidemic diseases can be dealt with much more cheaply
and effectively. An improvement in general housing conditions
has occurred in rural as well as urban areas. The spread of
corrugated metal and even scrap metal roofs as replacement for
thatch has helped to reduce problems of vermin in houses and

makes possible the collection of comparatively clean rain water for washing as well as consumption.

It is not as clear that the nutritional situation has improved. There are reports of loss of protein from the diet as increasing population densities lead to a reduction in the number of both wild and domestic animals and require the use of high yield but less nutritious crops such as cassava. On the other hand, the improvement in transportation and food storage facilities has meant that the most serious effects of food shortages can be avoided. The gradual disappearance of traditional food taboos is helping to bring about better consumption patterns, also.

Finally, the most important of these direct factors must be the increase in public knowledge of how to deal with health problems. During the past decade, following the attainment of political independence, there has been a very sharp rise in the proportion of children who receive formal education. In a number of countries, one of the largest single areas of investment in the national budget is the school system. Levels of literacy are rising and mass media, especially radio, are widely available for adult educational purposes. As a result, there is growing public awareness of basic concepts of hygiene and health care.

For the future course of mortality in the region, a continuing gradual decline appears plausible. Programs of public health and preventive medicine almost certainly will continue to be developed, although not always at the fastest possible pace. Moreover, without large steps in the control of major diseases such as malaria, the decline only can be gradual.

One barrier to any such breakthrough is that ministries of health are often not in a strong position to compete for scarce development funds. In some countries, direct expenditures on health care are barely keeping up with the growth of population. Following political independence, there has been an expansion in the number of medical schools (6 in 1961, 20 in 1968); as they begin producing graduates, the supply of qualified

physicians may be expected to increase. However, in some
instances the curricula are quite conservative and it is not
at all clear that the physicians will be trained in such a way
as to make the greatest possible contribution to reducing rural
mortality and morbidity. Moreover, in many countries a large
proportion of existing medical staff is expatriate. In Nigeria
in 1970, for example, just about half of all physicians were
non-Nigerian. It is likely, therefore, that for some time much
of the output of the medical schools will be absorbed simply
in the task of replacing foreign staff. There is also likely
to be some loss of African physicians to the brain drain.

Fertility

In Black Africa, the data are more than adequate to establish that fertility is among the highest in the world. Crude
birth rates are estimated to be typically between 45 and 50 per
thousand population. Indeed, in a number of countries, there
are indications that the rate is even higher than 50. By
contrast, in both Latin America and Asia birth rates are on the
average below 45. There are a few countries in the latter two
regions where birth rates reach approximately 50, but this is
becoming the exception. On the other hand, birth rates below
45 are distinctly uncommon in Black Africa.

In broad terms, fertility is higher in West Africa in the
belt from roughly Senegal to Nigeria and is also very high
throughout most of East Africa. A zone of lower fertility is
found centered around Gabon, Equatorial Guinea, and into parts
of Cameroon, the Congo and Zaire. Within this region of lower
fertility, crude birth rates are estimated to be something like
25% below those found in much of the rest of Sub-Saharan Africa.

A satisfactory explanation of fertility differentials will
require considerably more close attention than has heretofore
been given. Leading factors appear to be variations in levels
of health of the population, with venereal disease, chronic
malaria and malnutrition playing imprecisely known but probably
leading roles.

Unique cultural factors, such as variations in the age at
which sexual activity commonly begins, is also of considerable
importance. A good deal of attention has been devoted to the
question of the effects of the strict observation or the breakdown of a traditional postpartum taboo. Similarly, variation in
the average age of weaning has been asserted to be of considerable importance. Unfortunately, the relative importance of each
of these factors is unknown (Caldwell 1974; Morgan n.d.).

Some social variables such as marital instability, educational attainment and female participation in the labor force
have been shown to have an influence on fertility. However, the
aggregate demographic effects are often marginal. For example,
in Kenya in 1962 the census results indicated that African
women with nine or more years of formal education had some 15-20%
lower fertility than the average. The number of women with
that level of education was less than one percent of the total,
however (Kenya Population Census, 1966).

The evidence indicates quite clearly that conscious
attempts at control of fertility by means of contraception,
sterilization, or induced abortion have thus far not had any
noticeably significant demographic impact. In every national
society of Black Africa, undoubtedly some amount of family
planning by means of contraception is carried on. However, it
is mostly restricted to the highly educated, largely urban, more
affluent elite. All the data indicate quite clearly that consciously motivated effective limitation of fertility has not
reached very far into village life.

A useful body of comparative survey research exists concerning knowledge, attitudes, and practices associated with
family size in several countries of Black Africa. Although
such data must be interpreted with great caution, they help to
give a general picture of the social context of fertility in
the region. Essentially, the studies reveal that when asked
what they consider an ideal number of children, both men and

women respond with numbers on the average higher than elsewhere in the world (Caldwell, 1968). Commonly, respondents give ideal numbers of children which are not very different from the average actual completed family size. At the same time, almost all the surveys show that there is a significant minority of potential parents who say they want no more children.

A leading motivation among those who wish to control fertility is the cost of raising children--in particular, school fees. Desire for education is intense and all schools, public as well as private, charge tuition.

These surveys also have shown that knowledge of means to control fertility is very limited, although beginning to increase a little. Responses reveal a good deal of misinformation, and much of what is known involves methods that are ineffective or possibly dangerous. Of course, the surveys indicate very slight practice of contraception.

On the basis of the inquiries, there is little evidence of any very rigid moral or social barriers to the use of contraceptives. There is fear of sterility and child mortality, of course. Given existing health conditions, such fears are not difficult to understand. The concerns are all the more serious in societies where children are the chief means of support in old age, and in general the kin group is a major source of social and personal security. However, another very important factor supporting the persistence of high ideal numbers of children is almost certainly lack of information. In the absence of knowledge about the means to safely and reliably plan childbearing, it is unlikely that the ideal will be very different from the actual.

Within the region, three nations have programs of direct and active government intervention aimed at reduction of the rate of natural increase by means of family planning: Kenya, Ghana, and Botswana. Here, government funds and personnel are specifically devoted to making contraceptive information and supplies available to all segments of the population.

In the Kenya program, for the past year or so there have been approximately 4,000 contraceptive acceptors each month. This is a notable achievement, but it must be recognized that each year the program is reaching less than 2 percent of the total number of women in the childbearing ages. Quite obviously, even if every new acceptor were perfectly successful in controlling her fertility and if none were merely switching to the government program from private means of contraception--which is certainly not the case--the demographic impact would still be nearly imperceptible. The situation is essentially the same in Ghana, where the program has been serving about 2,500 acceptors per month.

In a number of other countries, the national governments permit the provision of family planning agencies such as municipal or local governments, medical service institutions, or by voluntary associations. Countries in this category include Liberia, Nigeria, Tanzania, Zaire, Mali, and others. Of course, there is a considerable range in the amount of enthusiasm with which the various central governments view family planning. In some, a national program exists in all but openly stated policy; in others, it is a matter of no more than bare governmental toleration of voluntary association activity.

There is another set of countries in which there is no official policy concerning family planning but where importation of contraceptives is difficult and where they are available only through private physicians, thus being effectively available only to the affluent.

There are also a few countries where family planning is actively discouraged. These include Malawi, Gabon, and Cameroon.

Some specific comments on the future course of fertility contron policies might be made. First, it can be plausibly argued that existing programs will have difficulty producing a significant impact on fertility if they continue along the present narrow lines. The medical institutions of the region

are not sufficiently staffed to mount an effective family planning program, especially given the excessive levels of
mortality and morbidity which they confront. It should be
remembered that the national family planning programs of Kenya
and Ghana have two of the best staffed medical institutions to
draw on; almost any new programs would have to contend with
medical institutions that are far less well equipped, operating
in societies where mortality is even higher. An implication is
that in order to increase the control that the people of the
region have over their own fertility, it will be necessary to
have methods that depend less heavily on the present medical
service institutions for delivery to the population.

Apart from the issue of the capability of countries to
implement a fertility control program, there are a variety of
factors associated with the decision as to whether or not there
should be such a policy.

One of the important factors affecting the importance
given to fertility in a country's population policy is the
existence of a reliable body of demographic data. Those
nations where population size and rate of growth are most
accurately known are also those most likely to view favorably
an active family planning program--Ghana under the leadership
of President Nkrumah being the most notable exception. Another
important factor is the residual influence of previous colonial
traditions, including organized religion. In the Anglophone
colonial countries, voluntary family planning associations (like
many other special purpose voluntary associations) were active;
in the Francophone countries they were virtually unknown.
Similarly, Protestant influence in the Anglophone colonial
countries was probably more favorable to open acceptance of
contraception than was the Roman Catholicism of the Francophone
countries. This has made no visible demographic difference but
it gave organized family planning a more immediate legitimacy
in the countries with the Anglophone experience which persists.

(It should be clearly noted that this may have no more to do with
the long run course of fertility in Africa than it did in Europe.)
An immediate factor is political stability and the extent of
social and economic planning; countries where the government can
only manage to stay in power on a day-to-day basis are unlikely to
be able to give much attention to any policies concerned with long
run basic social change. Finally, it is essential to observe that
there is no correlation between the political structure or the
socio-economic ideology of governments and their population policies regarding family planning.

Within the past few years, especially since the African Population Conference in Accra, December, 1971, and echoed again with
the World Population Conference in Bucharest, August, 1974, a
critical reevaluation of family planning as the core of population
policy has become more articulate and deserves close attention.
The critical view is not the specific policy position held by any
country in the region but it has attracted the interest of many
politicians, scientists, and planners (Okediji and Bahri, n.d.).

There is some variation among proponents of this view. It is
offered as the rationale for a blanket and sometimes almost hyster-
ical condemnation of all assistance in the area of population
policy and of all discussion of population issues in socio-economic
terms, and elements of the view have been used as a basis for a
reconsideration of population policies in order to achieve a
better understanding of how they relate to desired social and
economic goals.

A central theme is that the one essential business of policy
should be social and economic development reaching all segments of
the population (and not merely economic growth in the aggregate);
that unless programs of fertility reduction are carried out in the
context of such policies, they may well do more social harm than
good. A position sometimes associated with this theme is that the
motives of the foreign agencies urging family planning are funda-
mentally reprehensible; they are designed to maintain the existing

state of dependency of the less developed countries and protect
the privileged position and superconsumption of the developed,
especially Western, countries (Mauldin et al 1974). This critical
view has attracted particular attention in Black Africa because
of some special circumstances.

First, the proponents of strong and specialized family planning programs have little to show as successful cases within the
region. Favorable results in East Asia are not felt to have much
relevance to conditions in Africa; the tendency of some visiting
advisors to appear to assume that all developing countries are
pretty much alike is seen merely as insensitivity to the unique
African experience and potential. The small size and comparatively low density of many of the countries in Sub-Saharan Africa
are believed to offer opportunities that invalidate the applica-
bility of East Asian models.

Second, the critical view relates directly to the very
tenacious persistence of widespread poverty in the region after
more than a decade of striving for economic gain. It appears also
to speak meaningfully to the issue of a growing gap between rich
and poor embedded in the structure of many societies.

Third, the very single-minded dedication of some of the proponents of strong family planning programs at times has encouraged
the emergence of such a critical reaction. Population has sometimes been discussed as if it could be isolated as a matter of
excess fertility and nothing else.

All this has aroused some skepticism and suspicion of the
motives of those who advocate the development of family planning
programs. The extent to which the articulation of this point of
view will have an effect on government-level decisions concerning
population policy remains to be seen. However, it is likely that
elements of this view may have some deterrent effect on consideration of population policies in general and on the provision of
contraceptive technology to the public policies in the region.

Apart from the question of any government policies, there is
no evidence of downward movement in the fertility of any nation in
Black Africa. To the contrary, there is fragmentary evidence to
suggest that if there is any movement away from a high steady
level, it is a rise in fertility. In a particularly revealing
piece of research, Henin (1968, 1969) has shown that in the Sudan
the change to more productive agricultural techniques (from nomad-
ism to rain cultivation or to irrigation) is associated with a
rise in fertility. Moreover, in those countries with a series of
reasonably reliable censuses, the more recent data show that
fertility remains at a high level or is rising. For example, in
Tanzania the crude birth rate is estimated to have gone from about
46 to 48, between 1957 and 1967. In Kenya, the estimated crude
birth rate went from 48 to 50 between 1962 and 1969 (Blayo and
Blayo, n.d.; Henin and Egero 1972). Such small differences in
crude measures should not be given too much weight, but they do
support the proposition that fertility may tend to rise.

There probably are the beginnings of a fertility decline
through use of contraception and induced abortion in some urban
areas, especially among the better educated and more affluent.
However, no such trend is observable for any whole society. On
the other hand, as malaria and venereal disease are brought more
completely under control, or if there is a general shortening of
the period during which the postpartum taboo is observed or lacta-
tion is carried on, or if there is a fall in the age at which
regular sexual activity on the average begins, one may expect
some rise in fertility. For the immediate future, the realistic
expectation must be for continuing or slightly rising high fertility.

An important result of the high and generally steady levels
of fertility throughout Black Africa is that the populations are
on the average very young. Specifically, the percent under age 15
is almost everywhere in the 40's. One consequence of this is a
very high rate of age dependency. Another, and perhaps even more

important consequence, is that population growth has tremendous momentum behind it. The new parents of the next decade or two are already born and they are very numerous indeed. Even if fertility rates were to begin to decline substantially and immediately, the numbers of births can remain high for a long time, simply because the quantity of parents in the population is so great.

The most obvious effect of the high and stable or slightly rising fertility coupled with declining mortality, is that the populations of Black Africa are growing at a high and accelerating rate. Broadly for the region, the populations are now estimated to be growing at slightly over 2-1/2 percent per year. This is somewhat lower than the rate of increase for Latin America, where mortality has already reached low levels, but not very different from the rates of increase found in Asia. However, it appears that rates of increase may be tending to decline in Asia and Latin America but may very well rise in Africa. At the present rate of increase, most countries in Black Africa will double in population in just over 25 years if not sooner.

REFERENCES

African Census Program, 1973. "Report of the Secretary-General of the United Nations on the African Census Program." Seventeenth Session of the Population Commission, Geneva, 29 October-9 November (ms).

Amin, S. (ed.), 1974, Modern Migrations in West Africa, London: Oxford University Press.

Blayo, C. and Y. Blayo, n.d. "The Size and Structure of African Populations" in Population in African Development, P. Contrelle (ed.), Dolhain, Belgium: Ordina Editions.

Caldwell, J.C., 1968. "The Control of Family Size in Tropical Africa," Demography, 5, 598-619.

_____, 1969. African Rural-Urban Migration. New York: Columbia University Press.

_____, 1974. "The Study of Fertility and Fertility Change in Tropical Africa." Unpublished ms.

Faulkingham, R., and P. Thorbah, 1974. "The Demographic Impact of Drought: An Ecosystem Study of a Village in Niger," unpublished ms.

Gaisie, S.K., 1975. "Levels and Patterns of Infant and Child Mortality in Ghana," Demography, 12, 21-34.

Goldstein, S., 1973. "An Overview of World Urbanization, 1950-2000," in Internal Population Conference, Vol. I, Liege: International Union for the Scientific Study of Population.

Hance, W., 1970. Population Migration and Urbanization in Africa. New York: Columbia University Press.

Harris, J., and M. Today, 1970. "Migration, Unemployment and Development: A Two-Sector Analysis," American Economic Review, LX, 126-142.

Henin, R., 1968. "Fertility Differentials in the Sudan." Population Studies, XXII, No. 1.

_____, 1969. "The Patterns and Causes of Fertility Differentials in the Sudan," Population Studies, XXIII, No. 2.

Henin, R., and B. Egero, 1972. "The 1967 Population Census of Tanzania: A Demographic Analysis." Research Paper No. 19, Bureau of Resource Assessment and Land Use Planning, University of Dares Salaam.

Kenya Population Census (1962), 1966. Vol. III. Nairobi: Government Printer.

Mauldin, W.P. et al., 1974. "A Report on Bucharest." Studies in Family Planning, 5, No. 12.

Morgan, R.W., n.d. "Traditional Contraceptive Techniques in Nigeria," in Population in African Development, P. Contrelle (ed.), Dolhain, Belgium: Ordina Editions.

Okediji, F.O., and A. Bahri, n.d. "A New Approach to Population Research in Africa: Ideologies, Facts and Policies," in Population in African Development, P. Contrelle (ed.), Dolhain, Belgium: Ordina Editions.

Peil, M., 1974. "Ghana's Aliens." International Migration Review, VIII, 367-381.

Udo, R.K., 1968. "Population and Politics in Nigeria," in The Population of Tropical Africa, J.C. Caldwell and C. Okonjo (eds.), London: Longmans.

AFRICA AND THE PROBLEMS AND PARADOXES OF
FOOD PRODUCTION AND DISTRIBUTION

by

Harvey K. Flad

There is a basic paradox in the food situation in Africa
today: the need for more and better food supplies in an essentially agrarian society. The African population is the most
agricultural in the world; some 85% to 90% of the work force is
estimated to be employed in the agricultural sector. Yet this
same population is the poorest fed in the world. How has this
come about? Hunger, whether as a worldwide phenomenon or as a
regional problem, is the result of either insufficient production
or the maldistribution of a necessary resource--food. In Africa,
both causal factors are to blame; production in many areas is
poor, and many more areas are poorly serviced by distributional
systems.

Problems of Production

Viewing the recent droughts in Africa and the problems of
food production associated with them, climate has become a major
whipping post of observers of the agricultural scene. Although
the physical environment may not be the most important barrier,
it does play a significant role in placing constraints upon
development possibilities.

Aspects of the physical environment which impinge upon
agricultural production are continental, regional, and local in
nature. The continent's bulk (approximately 30 million square
kilometers, or 11,530,000 square miles), position (primarily
within the tropics of Cancer and Capricorn), and shape (little
configuration of the coastline) have all had an influence on
overall agricultural production and distribution patterns
(Bunting, 1970). The potential for internal exchange of commodities, for example, has not been realized, due, among other

reasons, to the vastness of distances to be overcome. Although temperature (particularly the length of the cold season) is not as inhibiting a factor in the tropics as it is in the temperate zone, the high rates of insolation and evapotranspiration, along with the length and severity of the dry season, set major environmental limits. This is often compounded because areas where the least rainfall is expected on the average per year are also the areas where the probabilities are lowest that any rain will fall at all. In addition, external trade relations for both land-locked countries as well as for most which have outlets to the sea have been hampered by the lack of adequate port facilities for ocean-going vessels.

The distribution of agriculturally useful soil types throughout Africa is quite limited. Soil formation is a function of parent material, time, climate, living organisms, and topography. The underlying geological structure of the continent has not developed soils which are extremely fertile or useful to agriculture. The Precambrian base complex granites have weathered over millions of years, often producing shallow clay pans which are seasonally inundated and generally difficult to work. Wind-blown loess-type soils on the margins of the Sahara are more friable, but lack sufficient structure for continuous cropping. Lateritic soils which are found throughout the heavily forested regions of the continent have residuals of iron and aluminum oxides but lack other nutrients for food crop production due to the leaching effect of tropical rainstorms. Also, high temperatures and heavy rains prevent the formation of a humus layer. Soils formed in regions of recent volcanic activity, however, are deep and fertile, and heavily used; the difficulty lies in the fact that many are found on steep slopes and are subject to erosion.

The hydrological pattern, like the pedological map, has great regional diversity and can cause problems in agricultural expansion. African river systems are vast, but have only recently been effectively tapped for irrigation. The subsaharan artesian system also has just recently been tapped, although it was extensively used during Roman times on the northern margin of the desert. In both cases the technologies for asserting control over the water supply have brought with them a host of problems. Irrigation projects, for example, have led to massive growth in water-associated diseases, and often have upset natural cycles maintaining soil fertility. Areas around the artesian wells in the Sahel have become overgrazed and subsequently heavily eroded.

Pests and diseases are another contributing factor to add to the list of environmental hazards in African food production. Grain supplies might be depleted by swarms of locusts or quelea birds; in 1958 167,000 tons of grain were destroyed by locusts in Ethiopia, an amount equal to a year's supply of cereal grains. Animal protein supplies are lessened due to cattle losses as a result of epidemics of rinderpest or trypanosomiasis (Knight, 1971: 43); it is estimated that 4.25 million square miles of good grazing land is unavailable for cattle production in Africa because of tsetse (McKelvey, 1973). Meanwhile, innoculation campaigns are a massive and expensive undertaking. Agricultural labor is often less productive due to diseases, such as river blindness (onchoceriasis), malaria, trypanosomiasis, schistosomiasis, or yellow fever. For example, the WHO (1966, p. 25) has noted for river blindness that

> In many parts of West and Equatorial Africa, more than 50 percent of the inhabitants are affected; 30% of them have impaired vision, and 4-10% are blind. In some villages of Upper Volta and in Ghana, the percentage of blindness reached 13-35%. Some parts of Nigeria are also seriously affected by onchocerciasis. Out of an estimated 350,000 victims of this disease, 20,000 are said to be blind. The average proportion of blind persons in the endemic areas is reported to be 10%, as against 0.2% in regions free from onchocerciasis. At Bamako, in Mali, two out of three persons examined suffer from onchocerciasis.

Such incidences are bound to affect the labor potential of the afflicted areas. The Aswan High Dam was built for irrigation purposes, as well as electric power production, and should have increased agricultural productivity; yet, the increase in incidence of schistosomiasis has drastically reduced the work efficiency of the farmers. There is concern that this problem may also accompany other irrigation schemes in tropical Africa, for example, the Kainji Dam in Nigeria.

Within the constraints imposed upon the potential for agriculture by the physical environment, the technologies of traditional and modern Africa have often promoted a cycle of environmental degradation. The literature dealing with African food production is most apt to list the traditional technique of slash-and-burn (shifting, or swidden) as the number one problem. As a given population increases, and as land becomes increasingly scarce, such extensive agricultural systems are less able to adapt to the mounting pressures. Indeed, there is a population "problem" in Africa (Hance: 1968 & 1970); one aspect of the food-population-land dilemma is the rapid growth of rural populations in regions of limited carrying capacity (Allan: 1965). Wherever populations increase beyond the capability of the local systems to support them, soil degradation, erosion, land fragmentation, and rural unemployment increase, and productivity per caput decreases. But this analysis of the problem seems somewhat linear; as I will argue later, this does not take into account traditional intensification adaptations that have occurred. Boserup (1965) might suggest that it is precisely this pressure of population growth which is necessary in order to promote change within a traditional framework.

A more important factor in regard to a decrease in the production of food crops is the stress on export production over subsistence production, an enduring legacy of the colonial system in Africa. In many situations the influence on food production was direct, as when a plantation economy was imposed

in an area of traditional subsistence food production. A result
of this activity was to take land (usually the best) out of the
production of foodstuffs; food crops often were and are relegated
to the poorer lands. On the other side of the coin has been the
change of diets accompanying urbanization. Diets of the affluent
particularly have changed to include non-African produced products. These can only be imported, thus leading to an important
source of capital outflow from developing agrarian economies.
This is an example of the paradox at a national scale of the
disparity of food consumption.

Both traditional and modern technologies have other problems which influence their ability to produce enough food for
the increasing population of Africa. One is the role of petroleum production and prices upon African food production; the
other is methods of storage.

The quadrupuling of oil prices in the past two years has
created a major bottleneck to increased food production in
Africa, as costs for fuel and fertilizer have spiralled.
Increased energy costs affect agricultural production at the
very base of change from traditional to modern technologies.
In advanced agricultural systems the energy input is enormous:
in the U.S. it takes the equivalent of 80 gallons of gasoline to
produce one acre of corn; or to put it another way, two calories
of fuel are required input for every calorie of food output
(Borgstrom: 1974, p. 18). Costs for oil and gasoline to drive
tractors, pumped wells, irrigation equipment, and engines for
milling and other manufacturing processes have grown so high
that many operations have had to cease. In some instances, such
as in the use of tractors perhaps, the return to more labor
intensive rather than capital intensive technology might be
welcomed; but, when the lack of fuel means a breakdown in the
sanitary water supply, or perhaps non-growth in the manufacturing
sector and a subsequent loss of jobs, the losses far outweigh
the gains. Increased petroleum prices have particularly affected

the cost and supply of fertilizer. Political scientists like
Kissinger and economists like Ewell claim that fertilizer is the
most important single factor necessary to increase food production. This is especially imperative in the developing nations
where soils and plant varieties produce higher yields with the
addition of extra dosages of nitrogen, phosphate and potash than
do American or European fields with the addition of the same
amount of fertilizer. However, to meet the need for fertilizer
it is estimated that one new plant (costing approximately $100
million) must be built for every additional 6 million people.
This is both an economic and a political issue, as Barraclough
(1975) points out, since the production and marketing of fertilizer is controlled by multinational corporations; and further,
as Grant (1974) notes, Americans are not about to change their
habits much:

> Meanwhile, as the world is caught in a critical
> shortage of fertilizer for food production, and
> as we restrict our exports of fertilizer and
> food, Americans are applying some three million
> tons of nutrients to lawns, gardens, cemeteries,
> and golf courses--more than used by all the
> farmers in India, and half again as much as the
> current shortage in developing countries.

Lastly, the rise in oil prices has severely affected the national
development plans of African countries. In only very few, such
as Libya and Nigeria, has the effect been positive. Increased
income to Nigeria is expected (according to the Third National
Development Plan, April 1975) to increase food crop production,
fertilizer and pesticide production, irrigation, and infrastruc-
tural development, as well as produce great expansion in the
industrial sector. For most African nations, the increased
price of oil has hampered food crop production. Ghana's
"Operation Feed Yourself" has been reported in jeopardy due to
petroleum imports rising from $53 million in 1973 to $200 million in 1974 (NYT 1/26/75; F-15). The report continued:
"More than 30 less-developed countries, most of them in Africa,

which already had mounting food deficits, have been classed by
the United Nations as the 'hardest hit' by oil costs. Oil producing countries, such as Nigeria and Indonesia, can absorb
their oil earnings for their own development." This is, then,
an example of the paradox of disparity at the international
scale within Africa, and, like diets, is more a problem of the
distribution of the fruits of the resource than of its production.

Much of the existing food production in Africa is lost
during storage. Estimates vary since statistics on farmstead
production, storage, and use are haphazard at best, but one
estimate by the F.A.O. (1969: 6-7) suggests that world storage
losses for cereals, pulses, and oilseeds were of the order of
10 percent (5 percent from insects and mites, 4 percent from
rodents, and 1 percent from mold fungi). Other estimates range
up to as much as 50 percent (Brody, 10/18/75). Data reported
from Africa show high losses in both quantity and quality of
cereal grains during storage. In northern Nigeria during 1962
unthreshed sorghum suffered a mean weight loss of 8 percent due
to insects; this was approximately equal to 115,000 tons, which
"could have been sufficient to satisfy the usual cereal requirements of 1.3 million people." (FAO, 1969: 8). Losses in nutrient value are even more difficult to assess; generally this
occurs by selective feeding on the germ by insects, or chemical
changes due to fungi in the humid tropics. Part of the problem
of storage is a result of the climatic variables: dry seasons
in much of the continent are extremely long, allowing plenty of
time for insect infestation or mold growth; in other areas the
excessive humidity of the wet season increases the chances for
fungal growth. Generally, traditional storage construction is
adequate to the task of protection from the weather, although
less adequate in preventing rodent damage, insect infestation,
or nutrient losses due to fungi (Morgan, 1959). Associated
with the physical problems of storage are those that deal with

the control and handling of the product. Local farmers are more apt to take care of their own subsistence foods in storage, but less apt to have the proper facilities. Regional governmental facilities are more likely to be composed of the better materials and have the necessary fungicides, but movement to and from the farms requires investment in all phases of the distribution infrastructure.

The social dimension of agricultural production has been indicated in much that has been already stated above. Traditional farming methods are more often a result of cultural persistence rather than ignorance of other methods; this is true for swidden agriculturalists as well as for transhumant nomads, for the use of the short handled hoe or the poorly constructed storage bin. Cultural persistence is particularly associated with diet, and supplementary sources of protein, say, are often found unacceptable. Changes in land tenure occasion the greatest social dissension. Such changes are commonly a result of either population increase and subsequent land shortage, or concommitant with a change to the cash economy and concepts of freehold rather than communalism. Food production can be adversely affected in both instances: in the first because the arable land becomes too fragmented for effective use; in the latter because cash cropping becomes more important than foodstuffs, or because communal grazing lands are placed under the plow (Bennett, 1969). In some cases, such as in Ethiopia, the land holdings of the oligarchy and theocracy were so vast and little used for general foodstuff production that the agricultural production of the entire country suffered. In far more cases, food crop production suffered as a result of a stress upon cash cropping.

Problems of Distribution

Differences in the distribution of enough food and enough of the right kinds of food produce different effects between ecological regions, between and within nation-states, between

socio-economic classes, and within families. Essentially, the
question revolves around nutritional deficiencies as a result
of inadequate supplies of the right foods or enough foods for
certain groups in the population at certain times.

Food shortages may be of two kinds: a lack of calories
results in undernourishment, whereas a diet deficient in any of
the essential nutrients (vitamins, minerals, fats, carbohydrates,
or especially proteins) leads to malnutrition. In cases of
extreme deprivation, protein-calorie malnutrition may occur.
Hunger is associated with the lack of sufficient energy input
in the form of calories. An average figure must obviously
adjust to differing requirements according to such factors as
body weight, climate, and activity level, but the United Nations
has given a world average per caput requirement of 2350 calories
per day (Table 1). Thirty-four African countries have average
caloric consumption rates less than that minimal world figure,
and only seven are over that figure; none approach the 3000
calories per person per day figures of the Developed World. On
a map of the world showing food deficit areas, 1958-9, most of
the African continent is shown as lacking an adequate supply of
calories, protein, or both. (Grigg, 1970, fig. 1.7). In
essence, then, Africa is a hungry continent.

African diets not only lack enough food, they also lack
the right kinds of food. Malnutrition is severe, particularly
among the most vulnerable groups of children and pregnant and
lactating women. Although the compilation of grams per day per
person of total protein suggests that there are sufficient
total supplies, nevertheless, needs vary greatly according to
body weight, age, activity level, and growth requirements. On
a map of the world indicating "infant malnutrition", the
African continent stands out as the most severely affected.
(FAO, 1970, p. 30). Indeed, it is the children of Africa who
suffer the most from dietary inadequacy:

TABLE I

AFRICA: PER CAPITA FOOD CONSUMPTION

Country	Year of data	#/day Calories	Grams/day Total Protein	Animal Protein
Algeria	64-66	1890	55.7	6.6
Angola	64-66	1910	39.9	9.4
Burundi	70	2330	61.0	4.9
Cameroon	65-66	2230	58.9	10.8
C. A. Rep.	64-66	2170	47.5	12.0
Chad	64-66	2240	78.4	13.8
Congo	64-66	2160	39.8	16.0
Dahomey	64-66	2170	52.2	8.0
Ethiopia	70	1980	66.3	11.0
Gabon	64-66	2180	51.0	27.2
Gambia	64-66	2320	62.2	14.6
Ghana	66-68	2070	43.0	7.3
Guinea	64-66	2060	45.4	6.0
Ivory Coast	64-66	2430	59.1	12.9
Kenya	70	2200	68.0	15.9
Liberia	64-66	2290	41.4	8.5
Madagascar	70	2240	51.2	12.6
Malawi	70	2400	63.1	5.3
Mali	64-66	2130	68.4	15.0
Mauritania	64-66	1990	73.4	37.5
Mauritius	70	2370	49.5	13.9
Morocco	64-66	2130	57.7	10.0
Mozambique	64-66	2130	40.4	4.6
Niger	64-66	2170	77.5	12.5
Nigeria	70	2290	59.9	8.4
Rhodesia	64-66	2550	73.2	14.4
Rwanda	64-66	1900	57.0	3.6
Senegal	64-66	2300	64.0	21.2
Sierra Leone	64-66	2160	49.2	9.1
Somalia	64-66	1770	56.9	22.2
*South Africa	64-66	2730	77.0	28.3
Tanzania	70	1700	42.5	15.4
Togo	64-66	2210	50.9	7.3
Tunisia	64-66	2200	62.9	10.9
Uganda	64-66	2160	55.9	15.1
Upper Volta	64-66	2060	70.3	5.3
Zaire	64-66	2040	32.7	8.9
Zambia	64-66	2250	69.4	10.7
Egypt	64	2770	79.9	10.5
Libya	70	2630	66.1	19.7
Sudan	64-66	2090	58.9	18.7

*incl. Botswana, Lesotho, Swaziland, and Namibia

Av. World Requirement		2350	30.0	
U.S.	70	3300	89.7	59.6

Source: Tables 136 & 137, pp. 442-455. FAO, Production Yearbook: 19
(Vol. 25), Rome: FAO, 1972.

> They are growing rapidly and need not only
> relatively more food than adults but also food
> of a considerably higher quality. In propor-
> tion to its weight a six-month-old infant needs
> about twice as many calories and about five times
> as much high quality protein as the average
> adult; a two-year-old needs about 70 percent
> more calories and about three times as much high
> quality protein; and a four-year-old, about 50
> percent more calories and about twice as much
> protein. (FAO, 1970, pp.8-9)

Severe forms of protein calorie malnutrition (PCM) such as
kwashiorkor or marasmus are endemic throughout Africa; "kwashiorkor" in fact is a word which means "disease that occurs when
displaced from the breast by another child." (Jelliffe, 1969,
p. 76). It is estimated that about 15 million children a year
die before the age of 5 of the combined effects of infection and
malnutrition; this would indicate a figure close to a quarter of
all deaths in the world (Schmeck, 1974, p. 1). Even when children survive these severe forms of PCM, morbidity and mortality
rates increase, and mental and physical development decreases.
(Eastern African Conference, 1969). For adults the situation
can be equally disastrous to health and labor productivity.
"This lack of productivity tends to be self-perpetuating. The
person who can work only a few hours a day can't earn enough to
buy the food that would make a longer work day possible."
(Schmeck, 1974, p. 42).

Sufficient quantity of protein is not enough, however, for
the supply must include all the essential amino acids (biological
value) and be easily accommodated into the human body (digesti-
bility). Animal proteins are most effective on both counts, and
African per capita intakes are very low in both grams of animal
protein per caput and as a percentage of total protein intake.
On the average, animal proteins account for 30% of the protein
intake: meat and poultry 15%; milk 11%; fish 4%; and eggs 2%.
In only six African countries is the world average reached. The
proper mixture of vegetable proteins can alleviate much of this

problem on a world-wide scale; oilseeds, pulses and grains can be combined in "complementary protein" dishes to give full quality.

At present there remains a distributional disparity between populations in forest regions who have diets composed largely of starchy roots and tubers (cassava, sweet potatoes and yams) and those in savanna regions who have diets of cereal grains and/or animal products which are higher in protein value. And there remains a disparity of protein distribution in families, between those that have the greatest need (children and pregnant women) and those who can command the choicest foods (men).

One of the traditional methods of overcoming certain nutritional deficiencies is the ingestion of earth. Earth-eating (or geophagy) has been documented throughout Africa (Hunter 1973). Chemical analysis of geophagical soils from Nigeria (Vermeer, 1966) and Ghana (Hunter, 1973) attest to the nutritive value of this practice, especially in adding precious quantities of iron, calcium, and certain trace elements. The practice of geophagy is especially important in the forest zones, where some societies are lactose-intolerant, and usually therefore deficient in calcium.

Hunger and its associated malnutrition is most severe during the early part of the rainy season when last year's food stocks are exhausted and physical activity levels for planting and weeding are highest (Hunter 1967). Times of harvest and plenty are associated with ceremonies and rituals, and act in rhythmic counterpoint to lean times. Seasonal hunger affects all members of the community. Hunter's data from northeastern Ghana showed that 94% of the adult males and females active in farming experienced weight loss, as did lesser percentages of pregnant and lactating women, children, and the elderly, and that there was some evidence of an increase in mortality and morbidity rates during the hungry season. In Islamic areas such losses of caloric input may also be accentuated by the month-long fast of Ramadan.

To some extent internal trade is generated because of the cyclic nature of this seasonal pattern of agricultural production and food requirements, of hunger and plenty. Livestock are sold in the savanna areas for cash to supplement dwindling food supplies and are shipped to the forest zones, while cereals and other foodstuffs flow into the seasonally food-deficient regions of the savanna. Thus, seasonal hunger acts to engage even the subsistence farmer into the system of cash for food.

Increasing disparities in the purchasing power of individuals in the developing nations of Africa is causing a severe dislocation of proteins and calories in the dietary regimes. The share of the poorest 40% of the population seldom exceeds 10%-15% of the national income; whether rural or urban poor, this population finds it difficult to afford an adequate diet. Efforts to build up an industrial state have caused underemployment in the rural sector, since many journey to urban centers for possible employment. This loss of labor at the farm level leads to less agricultural production and hence less food for the growing urban demands in many African countries, because there has been little mechanization of farming associated with the off-farm migration of the labor force. The problem is increased greatly when there is no chance for subsistence farming and the family is forced to purchase its food in the urban market; in such cases, poverty leads directly to malnutrition. Urban unemployment, especially of school leavers, promises to continue as urbanization is projected to double in this present decade. Government programs designed to aid the urban poor by price-fixing agricultural commodities often do not pay the farmer for increased costs of production associated with new inputs such as hired labor, machinery, fertilizer, pesticides, gasoline, or new irrigation techniques. In far too many situations, the diets of the urban poor continue to decline in quantity and quality, and even the standard of living of the majority of the population of Africa--those 80% or more who live in rural Africa and are "employed" in agriculture--continues to decline. Udo (1971: 426) has found that even the poor that live in "food-surplus areas" are "as badly off

in their caloric and protein intake as those in food-deficit
areas."

At the other end of the scale 7% of the population is
accruing 40% of the income:

> Thus the most disturbing problem of our times is
> the vast gap which is widening not between the
> developed and under-developed countries in the
> global sense but between the more and more numerous
> downtrodden masses, who constitute the majority
> of the Third World population, and a world minority
> which includes not only the majority of the people
> in the developed countries but also a minority of
> the people in the Third World. (Amin 1970, p. 30).

In other words, the food crisis in this respect is two-fold:
affluent countries consume most of the world's supply of high
quality food (60% of the U.S. grain output is consumed by cattle,
sheep, pigs, and poultry to provide animal protein to Americans,
while three fourths of the remaining 40% is exported as fodder
crops to bolster the meat industry of other developed countries);
meanwhile, elites within the developing world mirror this dietary
affluence and consume the lion's share of high quality foodstuffs
in their countries. In fact, the problem becomes even more
accentuated when it is realized that much of the national incomes
of these developing countries (which are controlled by the elites)
is spent on importing foods. This situation has led Rene Dumont
(1969, p.168) to conclude:

> The essence of Africa's problem is that the peasant
> does not feel committed, does not feel that his work
> furthers his own interests or those of his country...
> In Africa the minority causes such a drain on national
> revenue that it limits investment, destroys any
> chance of rapid development, and places economic
> independence in jeopardy.

Attempted Solutions to Problems of Production

Most of the literature on African agricultural development
deals with attempts to make changes in technologies in order to
lessen environmental risks, or in technologies designed to increase

yields per unit of land or manhour. Projects that have dealt with
the physical environment have included erosion control, irrigation
control, better storage facilities, mechanization, and fuller use
of natural fauna. Erosion control methods have included the construction of bunds, planting of permanent grasses and foliage,
strip cropping, changes in land use to only allow grazing, re-
siting of roads and paths, mulching of perennial crops, planting
of row crops on the countour, tie-ridging, and the construction
of basins to collect runoff. Water control schemes have generally
been associated with export crop production, such as cotton or
sugar cane, but foodstuff production for internal and external
markets also has been included. For instance, wet-rice production
from the Senegal and Gambia river basins, and the polder schemes
in and around Lake Chad that have been effective in promoting the
production of both rice and wheat. The construction of catchment
basins, earth dams, and hand-dug, artesian and pumped well systems
have been used throughout the continent to increase water supplies
for permanent agriculture. In northern Nigeria they have also
been used to create a more permanent meat supply in range management schemes which were effective in settling formerly trans-
human Fulani pastoralists. Having worked on such a scheme, I can
attest to the voluntary nature of the resettlement program; it
was particularly apparent to the herd owners when prices paid per
cow were much higher for the range-fed beef than for the tougher
and leaner animals that had been on the move for months. In
Senegal solar panels are being used to pump water to the surface,
making excellent use of the semi-arid atmospheric climate, and
alleviating the need for petroleum supplies. Information on new
methods of drying and storing produce, both root crops and cereal
grains, has been important in upgrading facilities both on the
farm and at the produce dealers' and millers' stores. Jameson
(1970) presents the data on Uganda, including recommendations
for the construction and upkeep of storage bins, and for insec-
ticides to be used to halt losses while in storage. Prospects

for more mechanized farming are a bit more clouded. Inappropri-
ate types and scales, as seen in Tanzania and Ghana, have led one
analyst to conclude:

> Unfortunately, all the experience of the past has
> provided warnings of difficulties, but few concrete
> guidelines for a more positive approach. In many
> cases, for instance, it is difficult to determine
> whether mechanization has failed because it was
> inherently uneconomic, or because it suffered from
> certain technical and managerial problems that
> could have been avoided or overcome. (de Wilde,
> Vol. I, 1967, p. 130).

Even the usually optimistic FAO concludes that "in African conditions animal traction, with suitable implements, is much more
widely applicable than electricity or the internal combustion
engine." (FAO, 1971, p. 113) Lastly, the fuller use of natural
fauna holds great promise as a source of animal protein. Fishing,
particularly of the fast growing Tilapia in artificial ponds on
individual farm holdings, in newly established reservoirs or
intensification of the industry in certain of the larger lakes
like Victoria or Rudolf holds out the most positive gain in the
short run. The "cropping" of wild animals, such as antelope, is
yet to be effectively organized on a sustained yield basis.

The major set of technological solutions to the problems of
food quantity and quality might well be examined under the rubric
of the "Green Revolution." Basically, the idea is to create new
varieties of food crops which have higher yield potential, thus
giving higher yields per hectare and per manhour. During 1972/73
over 90 million acres of cropland in Asia and North Africa were
planted to high yield varieties (HYV) of wheat and rice, and produced from 50%-100% more wheat and 10%-25% more rice in these
areas, or an additional half ton of rice per hectare on the
average (Jennings: 1974). The "revolution" has not reached most
of Africa south of the Sahara, and if and when it does there
promises to be a host of problems associated with its implementa-
tion. As a monocultural system, HYV's are generally more

susceptible to diseases and pests since they are often too
genetically "pure" and don't have a built-in resistance to local
diseases; also, they are usually planted so closely together that
pests can cause a greater amount of damage in a shorter amount of
time than in plots where a variety of crops are intercropped with
each other. And, although they make more "effective" use of
fertilization and pest control, they also require more to make
the system operative. Thus, HYV's increase the demand for high
energy inputs like fertilizer, insecticide, and mechanization at
a time when petroleum prices are skyrocketing beyond the control
of the small farmer. This fact, among others, tends to favor
the large farmer who has available land and capital to invest
and hence, the disparity grows between the small subsistence
farmer and the larger farmer.

Whether farmers use the new HYV's or not, there is increased
demand for nutrients to make African soils more fertile. Nitrogen
is an important requirement, and potash deposits exist throughout
the continent. Nigeria's natural gas supplies can be used to
produce needed fertilizers, rather than being burned off in the
oil fields as at present. Lastly, great potential continues to
exist in improving traditional manuring techniques, using both
animal wastes and green manures.

Improvements also can be made in growing foodstuffs which
yield higher vegetable proteins. New cereal varieties, such as
high-lysine corn and sorghums are easily adaptable to African
growing conditions and diets. A naturally occurring variety of
sorghum which has a third more protein and twice as much lysine
as other sorghum strains has recently been discovered growing in
Ethiopian fields. (Brody, NYT, 10/11/74, p. 78) Also, millets
(both Pennisetum and Eleusine spp.) are reported to have higher
nutritive value than either maize or sorghum. Pulses and oil
seeds have the highest potential for increasing protein in veg-
etarian diets. Groundnuts (peanuts) are already grown on a wide
scale throughout Africa for both internal consumption and

international trade. Cowpeas are a traditional source in some
areas and could be more widely grown. Soybeans, which are
extremely productive sources of vegetable protein are not a
major crop in Africa, although field tests indicate that they
can grow effectively and taste tests suggest a possible
breakthrough into some diets. (Some examples of its use include
soya flour-based breads and drinks.)

Since the basic technologies to improve yields and nutritive value of agricultural foodstuffs seem to exist, a fundamental
question seems to be "will the African farmer adopt these varieties or methods?" Although the usual answer in western agricultural extension literature suggests that the traditional
African farmer is the bottleneck, in fact a look at the history
of agricultural innovation in Africa persuades me that the
opposite is the case: African farmers will change their crops
sown and their methods of cultivation if they perceive it advantageous to do so. A massive collection of data from the Congo
Basin area analysed by Miracle (1967) supports this view. Also,
if one looks at the major staple food crops of Africa today
(maize, cassava, sweet potatoes, plantains), and realizes that
they came originally from the Americas or Asia, then one can see
that great changes have already occurred in dietary regimes: by
1958 these "foreign imports" accounted for more than half the
total calories in Africa south of the Sahara (Johnston: 1958)
and by the 1970's rice and wheat had joined the list of important
exotic food sources in Africa south of the Sahara. Further support for this perspective comes from intensive case studies
analyzing the causes of adoption of agricultural innovation; for
example, the change to growing maize by the Sandawe of Tanzania
(Newman: 1970), the expansion of cassava cultivation (introduced
in the seventeenth century) at the expense of yam (the indigenous
staple) in the forest zone of Nigeria (Uzozie, 1971), or the
development of indigenous terracing and intensive land use
systems on Lake Victoria's Ukara Island (Ruthenberg: 1971,

p. 118). Furthermore, both Newman (1970) and Brooke (1967)
attest to the significance of the famine experience in facili-
tating such change.

In general, African peasant farmers attempt to minimize
risks rather than maximize profits; therefore, innovations which
strike at the problems of production stability should be readily
acceptable. An important concern in some of Africa is land, and
the recent land reform act in Ethiopia whereby twenty-five acres
of arable land is to be given to each farmer is a major step
forward. More importantly, it should be recognized by governments and international agencies that incentives for growth in
the agricultural sector should be based primarily on prospects
for food self sufficiency, and secondarily on developments for
the export market. Once security has been established, economic
pragmatism can play its part.

Abercrombie (1967) points to the on-going transition from
subsistence to market agriculture in Africa south of the Sahara,
and maintains that to increase the potential for these changes,
governmental programs should focus on structural developments
such as transport facilities and marketing channels. This would
strengthen the production side of the African agricultural
dilemma, but would not insure that production will be based on
foodstuffs or that the rural populations will have greater access
to this production. As Cunningham (1973) notes from his research
in Tanzania, there must be a policy change that places emphasis
on the peasant farmer, and on the distribution of the fruits of
the increased production to these farmers.

Solutions Involving Distribution

The development of a more dense communications network, of
all-weather roads to move small-holder production to market and
of informational systems such as agricultural extension services
which will promote innovations, are major priorities in developing
the agricultural sector. So, too, is the establishment of

marketing boards. But, the adoption of innovation requires also a growth in the social environment of the small-holder community. In particular, the need for nutritional improvement is not only to combat hunger itself, but is also an extremely important input into the development process.

Investment in nutrition contributes both short and long term advantages to economic development (Cole: 1971). In the short-term, adults will have increased stamina and energy which can be used directly to increase agricultural production. In the mid-term, infant malnutrition control will lead to an increase in the ability of the new generations to innovate and become productive farmers, as their literacy, disease resistance, and physical abilities improve. And in the long-term, the decrease in infant morbidity may lead to more acceptance of population control ideas, or a "demographic transition" which is both natural and acceptable.

More equitable distribution of nutritionally important foods can play an important role in furthering economic development throughout agrarian Africa. Equally important, African countries must reform their social systems so that there is a more equitable distribution of food and capital among the classes. In order to correct the paradox of importing foods into agrarian economies, there must be systemic and ethical decisions made to give unto all their daily bread. Then, Africa can feed herself, and her exports can help to feed the world.

REFERENCES

Abercrombie, K.C. (1967) "The Transition from Subsistence to Market Agriculture in Africa South of the Sahara," in Readings in the Applied Economics of Africa, vol. 1: micro-economics. edited by Edith H. Whetham and Jean I. Currie. London: Cambridge University Press, pp. 1-11.

Allan, W. (1965) The African Husbandman. Edinburgh: Oliver and Boyd.

Amir, Samir (1973) "African Paradox: food imports to an agricultural continent have to go out with the 1970's," CERES, vol. 6, no. 4, pp. 29-31.

Barraclough, Geoffrey (1975) "The Great World Crisis I", The New York Review of Books, vol. 21, nos. 21 & 22, pp. 20-29.

Bennett, Charles J. (1969) "Economic Development and Social Change: some experiences of the Nandi and Kipsigis peoples of Kenya, 1900-1949," Unpublished manuscript, Dept. of Geography, Syracuse University, January, 1969.

Bird, David (1975) "A Hungry World Struggles for More Food," The New York Times, January 26, pp. 85-86.

Borgstrom, George (1974) "The Price of a Tractor," CERES, vol. 7, no. 6, pp. 16-19.

Boserup, Ester (1965) The Conditions of Agricultural Growth. Chicago: Aldine.

Brody, Jane E. (1974) "Experts for Pest Control to Increase World's Food," The New York Times, Nov. 28, pp. 1 & 23.

Brody, Jane E. (1974) "Search for Protein Crucial in Struggle Against Hunger," The New York Times, Oct. 11, pp. 41 & 78.

Brooke, Clarke (1967) "Types of Food Shortages in Tanzania," Geographical Review, vol. 57, no. 3, pp. 333-357.

Bunting, A.H. (1970) "Research and Food Production in Africa," in Research for the World Food Crisis. edited by Daniel G. Aldrich, pp. 31-52. Washington, D.C.: American Association for the Advancement of Science.

80

Cole, William E. (1971) "Investment in Nutrition as a Factor in the Economic Growth of Developing Countries," Land Economics, vol. XLVII, pp. 139-149.

de Castro, Josue (1952) The Geography of Hunger. Boston: Little, Brown & Co.

de Wilde, John C., ed. (1967) Experiences with Agricultural Development in Tropical Africa, vols. I & II, Baltimore: John Hopkins Pr.

DuBois, Victor D. (1973) "The Drought in West Africa - Part I: Evolution, Causes, and Physical Consequences" (West African Series). American Universities Field Staff Reports, vol. XV.

DuBois, Victor D. (1974) "The Drought in Niger - Part I: The Physical and Economic Consequences" (West African Series). American Universities Field Staff Reports, vol. XV.

Dumont, Rene. (1961) False Start in Africa. N.Y.: Praeger.

Dumont, Rene and Bernard Rossier. (1969) The Hunger Future. N.Y.: Praeger.

Eastern African Conference on Nutrition and Child Feeding Proceedings. (1969) U.S. Agency for International Development, Washington, D.C.

F.A.O. (1969) Bigger Crops - and Better Storage: the role of storage in world supplies. (World Food Problems No. 9) Rome: UN FAO.

F.A.O. (1970) Lives in Peril: protein and the child. (World Food Problems No. 12) Rome: fao.

F.A.O. (1957) Man and Hunger. (World Food Problems No. 2) Rome: FAO.

F.A.O. (1964) Protein: at the heart of the world food problem. (World Food Problems, No. 5) Rome: FAO.

F.A.O. (1971) The State of Food and Agriculture 1971. Rome: FAO.

Ghai, Dharam. (1973) "Reaching the lower 40 percent," CERES, vol. 6, no. 4, pp. 45-49.

Grant, James P. (1974) "While We Fertilize Golf Courses," The New York Times, August 28.

Grigg, David (1970) The Harsh Lands. New York: St. Martin's.

Hance, William A. (1970) Population, Migration, and Urbaniza-
    tion in Africa. N.Y.: Columbia U.P.

Hance, William A. (1968) "The Race Between Population and
    Resources", Africa Report, vol. 13, pp. 6-12.

Hunter, John M. (1973) "Geophagy in Africa and in the United
    States: a culture-nutrition hypothesis". Geographical
    Review, vol. 63, pp. 170-195.

Hunter, John M. (1967) "Seasonal Hunger in a part of the
    West African Savanna: a survey of bodyweights in Nongode,
    North-East Ghana." Institute of British Geographers
    Transactions, no. 41, pp. 167-185.

Jameson, J.D., ed. (1970) Agriculture in Uganda, second
    edition. London: Oxford University Press.

Jelliffe, Derrick B. (1969) Child Nutrition in Developing
    Countries. U.S. Agency for International Development,
    Washington, D.C.

Jennings, Peter R. (1974) "Rice Breeding and World Food Pro-
    duction", Science, vol. 186, no. 4169, pp. 1085-1088.

Johnson, Thomas A. (1974) "The 'Traditional Farmer' is Africa's
    Hope", The New York Times, Dec. 16.

Johnston, Bruce F. (1958) The Staple Food Economics of Western
    Tropical Africa. Stanford Univ. Press.

Knight, C. Gregory. (1971) "The Ecology of African Sleeping
    Sickness", Annals of the Association of American Geog-
    raphers, vol. 61, pp. 23-44.

May, Jacques M. (1974) "The Geography of Malnutrition in
    Africa South of the Sahara", Focus, vol. 25, no. 1 & 2,
    pp. 1-10.

McElheny, Victor K. (1974) "Rising World Fertilizer Scarcity
    Threatens Famine for Millions", The New York Times,
    Sept. 1, pp. 1 & 34.

McKelvey, John J., Jr. (1973) Man Against Tsetse. Ithaca:
    Cornell Univ. Press.

Miller, Norman N. (1974) "Journey in a Forgotten Land -
    Part I: Food and Drought in the Ethiopia-Kenya Border
    Lands" (Northeast African Series). American Universities
    Field Staff Reports, vol. XIX.

Miracle, Marvin P. (1967) Agriculture in the Congo Basin. Madison: Univ. of Wisconsin Press.

Morgan, W.B. (1959) "The Distribution of Food-Crop Storage Methods in Nigeria", Journal of Tropical Geography, vol. 13, pp. 58-64.

Newman, James L. (1970) The Ecological Basis for Subsistence Change Among the Sandawe of Tanzania. Washington: National Academy of Sciences.

Newman, James L., ed. (1975) Drought, Famine and Population Movements in Africa. (Foreign and Comparative Studies/ Eastern Africa, No.XVIII) Syracuse, N.Y., Syracuse University.

Schmeck, Harold M., Jr. (1974) "Malnutrition is up sharply among world's children", The New York Times, Nov. 5, p. 1.

Udo, Reuban K. (1971) "Food-Deficit Areas of Nigeria", Geographical Review, vol. 61, pp. 415-430.

Uzozie, L.C. (1971) "Patterns of Crop Combination in the Three Eastern States of Nigeria", Journal of Tropical Geography, vol. 33, pp. 62-72.

Vermeer, Donald E. (1966) "Geophagy among the Tiv of Nigeria", Annals of the Association of American Geographers, vol. 56, pp. 197-204.

Wade, Nicholas (1974) "Green Revolution (I): a just technology, often unjust in use." Science, vol. 186, no. 4169, pp. 1093-1096.

Wade, Nicholas (1974) "Green Revolution (II): problems of adapting a western technology." Science, vol. 186, no. 4170, pp. 1186-1192.

W.H.O. (1966) Committee on Onchocerciasis. Second Report, No. 335. Geneva.

CABINET MINISTERS OR CABINET MAKERS?

A Critical View of "Nonformal"
Education for the African

by

Charles H. Lyons

The past decade has witnessed a criticism of formal
schooling unprecedented in world history. I need not dwell
upon the recent global phenomenon of student strikes and demon-
strations which shared as a common denominator a criticism of the
university system and indeed formal schooling in general; nor on
the numerous journalists and social activists of the late 1960's
who continually produced works critical of the conduct and goals
of formal schooling. Initially their barbs were aimed at the
inferior forms of education offered to the poor and minorities,
but by the 1970's the criticism had developed to be a broad
indictment of what in many circles was portrayed to be a rigid,
mind-destroying, irrelevant educational system. The past decade
has also been one in which those connected with the study of
education have attempted to define education in a more useful
fashion. In part, the quest for redefinition was a response to
the critical spirit of contemporary times; in part, it was the
logical extension of a continuing debate over the goals and purposes of education which would probably have taken place anyway.
The quest for that new definition has had at its core an attempt
to portray education as something much broader than formal
schooling. Since these twin currents of criticism on the one
hand and redefinition on the other gave birth to the nonformal
education approach, we might with profit spend some time examining
recent educational history.

One might date the beginning of contemporary criticism
of the school with the publication, in 1966, of the now-famous
Coleman Report. Previously, it was generally assumed among

scholars as well as the public that the fundamental role of
schooling in modern democratic societies was to promote individual
and group mobility. It logically followed that, if a society
wished to ensure such mobility it would have to equalize educational opportunities. In the United States, where Coleman did
his research, the liberal argument ran that blacks and other
minorities were under-represented in high status competitions
because the schools they attended were of poor quality. Thus, if
you wished to ensure equality of educational opportunity for
minorities, it was necessary to improve the educational programs
of their schools. This was the fundamental assumption made by
the educational planners of President Johnson's "Great Society"
era. To prepare for its Congressional lobbying, the Department
of Health, Education and Welfare hired James Coleman in 1964 to
document just how inferior schooling for minorities actually was.
But as Coleman's computers spat out their results during the
Spring of 1965, it readily became apparent that the facts would
not lend support to the lobbying effort; indeed, they undermined
the basic assumption upon which the liberal argument rested.
(Grant 1973 and Hodgson 1973)

The Coleman survey--the most thorough ever conducted in
the United States--found that the educational opportunities provided minorities were just about on a par with those provided for
the majority. As if that were not surprising enough, it went on
to indicate that, when weighed against other factors (most notable
those connected with the family), the school's power to improve
the socio-economic status of groups and individuals was quite
negligible. This was a sorry finding for advocates of public
education who for years had pried open the public coffers with the
argument that schooling performed a valuable role in "democratizing"
and "equalizing" American society. Since the publication of the
Coleman report, social science investigation has not so much chal-
lenged as elaborated upon these findings. Prominent scholars
such as Daniel Patrick Moynihan and Christopher Jenks in the

United States and Raymond Boudon in Europe have continually
driven home the message that the school's role in promoting
greater socio-economic opportunity is minimal at best.

   This crisis of faith in the public school ideal led
inevitably to a search for the real purpose of formal education.
It rapidly became the vogue for scholars interested in education
to take the highly revisionist position that schools were not set
up to provide equal opportunity (that was just the rhetorical
pap used to sell the idea); rather they were established to train
workers in the practical skills and habits of subservience
needed by the modern corporate state. As the argument went,
schooling trains youths to cope with the bureaucratic structures
which they will later encounter in the world of work. The
school bell is an introduction to the time clock; the teacher, a
preparation for the foreman. Thus, the social attitudes and
sensibilities which schools consciously and unconsciously foster
aim at promoting in the young an unquestioning acceptance of the
dominant value system of modern industrialized society. If this
seems akin to a Marxist analysis of the role of education in
capitalist societies you are right, but with one caveat. The
position is less Marxist than semi-Marxist for, with only a few
exceptions, the critics seldom follow their analyses to the
logical conclusions. While they couch their criticisms in terms
of material power and resultant class conflict, they generally
fail to reach the logical Marxist termination point: a description of a socialistic, classless society. The strong points of
such works lies in their criticism; their weaknesses, in their
proposals for alternatives. (Karier, Violas and Spring, 1973;
Carony 1972; 1974).

   Let us consider, for one example, Michael B. Katz. His
The Irony of School Reform: Educational Innovation in Mid-
Nineteenth Century Massachusetts (1968) and Class, Bureaucracy
and the Public Schools (1971) represent classic expressions of
the semi-Marxist positions. Katz' first book, originally his

doctoral dissertation at Harvard, argues against the popular
interpretation that the foundation of public schooling in
Massachusetts was laid by a coalition of enlightened intellectuals
and working people who did battle with wealthy, conservative
factions. Instead, he argued that public schooling was in fact
instituted by the moneyed elite--the factory owners of Lowell
and Lawrence and the like--who sought to create a compliant,
obedient work force. The second book argues pretty much the same
position but, in more general terms. What hard lessons are we to
learn from all of this? One message of his earlier book is that
future educational reforms ought not be "too bound up with personal and group interests. . ." A message of the later book is
that schools should stick to teaching basic skills (the three
R's) and avoid influencing the social or political attitudes of
children. "People have a right to hold whatever beliefs and
whatever attitudes they please," writes Katz, "that is the only
consistent position for a civil libertarian. It follows that the
attempt to teach patriotism, conventional morality, or even its
opposite in a compulsory public institution represents a gross
violation of civil rights." Noble sentiments, perhaps, but in
the real world how might one avoid these pitfalls? In short,
Katz, while providing us with a provocative analysis, lacks an
over-arching world-view in which to interpret fully the material
he presents.

A similar comment might be made of Ivan Illich. Illich
is a true man of the world. European by birth and education, he
came to the United States in the early 1950's to work among the
Spanish-speaking immigrants of Manhattan. Later, he moved on to
the Catholic University of Puerto Rico where he became involved
in the training of North American missionaries for service in
Latin America. It was while he was in Puerto Rico that he became
very concerned about the mechanisms of human oppression and
gradually to believe that the educational system played a key
role in that oppression. In 1970, he codified his position in

De-Schooling Society, a book which caught the educational imagnation of men the world over.

He started with the premise that formal institutions of education are maintained by powerful elites in order to legitimize their privileged positions in society. An individual or two from a poor background may make it to the top of the educational ladder and reap the benefits which follow; there must be such examples if the masses are to accept the legitimacy of the educational system. In most cases, however, it is the sons and daughters of the elite who succeed educationally. They then wave their diplomas at the oppressed masses and assert that they are privileged, not because they were born to but rather because they "merit" their positions. In a fascinating fashion, Illich likens the role of the school to that of the Church of earlier times. In the pre-modern era, the Church held enormous secular power by virtue of its complete monopoly over the instruments of salvation. Eventually, however, the worldly power of the Church grew to be so dysfunctional that it had to be, in modern times, "disestablished." But as the Church became disestablished, the school became established; as earlier one needed the approbation of the Church in order to wield secular power, so today one needs the approval of the school system. Diplomas, degrees, and other credentials are the sacraments of today; without them, one cannot achieve worldly salvation. But, as in earlier times the monopolistic power of the Church grew to such an extent that it had to be disestablished, so in today's world the monopoly of the school is so great that it too ought to suffer the same fate. The equitable society is the "deschooled" society, that is, a society in which social benefits are distributed without regard to pieces of paper, a society which does not differentiate on such artificial bases as grades and diplomas.

As with the semi-Marxists, Illich proves long on provocative analysis of existing systems and short on an overarching social theory of what ought to be. When pressed, Illich does provide hints and suggestions of what the "de-schooled" society might look like and how it could be brought into being. Instead of schools, he maintains, men can be educated through such things as "skill exchanges" in which individuals knowledgeable of one field may teach others in exchange for their skills. Education could be gathered from "educators-at-large," men who hang out shingles professing to teach skills in exchange for financial reward; such "educators-at-large" would survive in the market place only if their teaching was successful. The world of the future would be one in which the monopoly of the school would be broken, in which real learning would take the place of mere credentialing. Further, Illich offers the happy advice to the developing world that they are in an advantageous position: whereas the developed world has the chore of dismantling its existing school system, the developing world can avoid this arduous step by not setting up a school system in the first place.

Beyond these few observations and suggestions we really get no sense of the type of social order which Illich seeks to establish. His is a romantic notion which calls to mind more the ambiguities of Rousseau than the ideology of a careful social theoretician. His view of the future is really a look backwards, towards a happy mythical past that never was. His proposals harken back to the apprenticeship patterns of traditional societies from which our present credential-conscious school systems evolved. If his fundamental aim is not so much to "deschool" as to do away with oppression, then he begs the question as to how we might prevent the growth in the future of other unfair mechanisms of social differentiation. As the school has taken over from the Church, so in the future some other hitherto unthought of mechanism might one day replace the school. Like Rousseau, he has faith that the human condition might be changed by tinkering with existing social institutions; others might postulate that the world is now too complex.

One thing which recent writers from Coleman to Illich
have accomplished is to spur educators to redefine what they are
about. Ever since the emergence of the modern teaching profession, those involved in education have tended to define education as schooling. The ancients' conception of education was
much broader. (Jaeger 1945). They had formal schools of
various kinds, but they were quick to stress that there were a
variety of institutions in society which educate: the family,
the religious cult, the army and so on. Similarly, the western
world up until the nineteenth century saw as important educational agencies the church, the community, and the family as
well as the school. However with the growth of the modern pub-
licly-supported school system and the consequent specialized
training of personnel to staff that system, the tendency developed to define education as formal schooling. In part this
tendency may be explained as an understandable if regrettable
effort of an insecure profession to make itself appear more
important than it is --and what could be more important than to
say that school teachers were responsible for the total education of society? In part, the tendency to equate education with
schooling also might be explained by the fact that, for a long
while, many of the educational functions of other social agencies
were taken over by the school. In early America, for instance,
the school took over the function of teaching reading precisely
because parents were failing in their traditional task of doing
so. In more modern times, the school has taken over other educational functions--ranging from sex to driver education--again
because other educational agencies and the family are failing in
their social responsibilities.

The fact remains, however, that much education goes on
outside the school. Parents and peer groups, the press and the
media, the church and synagogue and mosque do perform vital
educational functions. And, within the past decade or so, those
connected with the field of education gradually have come around

once again to see the totality of the educational enterprise.
They have done so not so much to downgrade the importance of
formal schooling as to understand the role of the school within
the wider educational context.

In the field of history and education, for instance, this
new view of education has resulted in a spate of revisionist
writing. It began, perhaps, with a small book by Bernard Bailyn,
Education in the Forming of American Society (1960). Bailyn
wrote this book when, early in his career, he was asked to teach
the quite unpopular course on the history of American education
offered by the Harvard Graduate School of Education. In his book
he was critical of the standard secondary accounts of the field
for committing three major sins of historiography: presentism,
institutionalism, and moralism. By presentism, Bailyn meant
that writers of educational history sought to force their concep-
tions of presently existing forms of schools back into history,
thereby distorting the past. By institutionalism, he meant that
educational historians, who were generally employed by the very
system about which they wrote, had too much at stake to write with
a sufficient sublimation of bias. By moralism, he meant that the
writers sought not to look into history in order to find out what
actually happened, but rather to derive, albeit in an artificial
fashion, moral lessons with which to exhort future teachers to
higher professional dedication. Bailyn argued that educational
history was generally bad history and that the subject was in
serious need of scholarly revision.

The revision he proposed had at its hub the placing of
the history of education within the wider context of social,
intellectual, and economic history. Education, he maintained,
was broader than schooling; hence, the scholarly educational his-
torian was one who looked at the myriad of historical forces and
currents which contribute to education. His own particular
interest was in investigating the educational role of families
and churches, but he opened the door for others to investigate

such agencies as the media, the military, the political process, and so on.

Shortly after writing his provocative essay, Bailyn returned to his first interest, the study of colonial history, and he has left for others the task of reinterpreting the history of education within the wider social context. Perhaps the most stimulating extension of Bailyn's ideas is Lawrence A. Cremin's projected three-volume history of education in the United States, the first volume of which came out in 1970. Unabashedly a convert to Bailyn's notion, Cremin has provided us with a most comprehensive definition of education. "Throughout the work," wrote Cremin, "I shall view education as the deliberate, systematic, and sustained effort to transmit or evoke knowledge, attitudes, values, skills, and sensibilities, a process which is more limited than what the anthropologist would term enculturation or the sociologist, socialization, though obviously inclusive of some of the single elements. Education, thus defined, clearly produces outcomes in the lives of individuals, many of them discernable, though other phenomena, varying from politics to commerce to technology to earthquakes, may prove more influential at particular times and in particular instances." One may disagree with this definition as there are a number of things which one learns not through any systematic, deliberate plan but rather through unconscious action or accidental happenstance. If one is struck by an automobile, for instance, one quickly "learns" about the adequacy of health care in his community, about management of hospitals, about the machinations of insurance companies. But this is not education in the restricted sense of Cremin's definition, as the victim of our accident did not step into the street to learn such things nor did the driver of the car run him down in order to "teach" him about health care. Cremin restricts his definition to the deliberate, systematic and sustained efforts to educate because it is precisely those efforts which men may control through conscious policy determination.

At the heart of Cremin's definition lies the notion of a curriculum. A curriculum is a deliberate, systematic, sustained effort to evoke knowledge, attitudes, and sensibilities. Mistakenly, we tend to think of curricula only in terms of schools with their lesson plans, course outlines, and so on. But, advertising agencies have curricula, as do newspapers in their editorial policies, ministers in their pulpits, and parents in their disciplinary procedures. A curriculum has a form, a plan of action, an ultimate goal. Thus, asserts Cremin, those concerned with national educational policy should think beyond the curriculum of the school and consider as well the curricula of numerous other educational agencies. He argues that schools of education should in the future no longer confine themselves to the preparation of teachers for the public schools; instead, they should also study the educational effects of the media, the family, etc., and they should begin to train people for the new professions in education which will undoubtedly emerge as a result of this more expansive view of education (Cremin, 1971).

As exciting as this is to us in schools of education, I would be remiss if I did not note that the Cremin position has had its critics. Sociologists and anthropologists may well suggest that there are a number of things one learns in the course of a lifetime which are not the result of anyone's deliberate, systematic plan. Cremin would counter, perhaps, by arguing that his attempt is to restrict his definition to those things which may affect men through conscious policy decision. The question remains as to how conscious a plan must be before it can be considered educational in Cremin's sense. Another criticism is that the Cremin definition is more "teacher-centered" than "learner-centered." The most serious criticism, however, might be that the definition is mechanical; it lacks a normative sense of what ought to be. His definition is just as applicable to

Nazi Germany as to classic Athens, to Stalinist Russia as to modern America. One needs more--a clear sense of social direction.

More hopeful, perhaps, has been Charles Silberman, the journalist whose Crisis in the Classroom became an instant bestseller when it was published in 1970. Silberman had been asked by the Carnegie Corporation of New York to prepare a report on teacher education in the United States. As he conducted his research, however, Silberman found that he could not do that without investigating the overall state of education, and that investigation in turn led him to consider the new, broader definition of education. Thus, he began his report with a highly expansive definition of education. He went back to the ancient Greek notion of paideia, a word impossible to translate precisely but which means, loosely, the panoply of educating forces in society. More than that, it also contains the notion of an ideal to which those forces should educate. The problem with American education, Silberman said, was not that the paideia did not educate, but rather that it educated without a comprehensive sense of purpose. Contemporary American education is, in his word, "mindless."

It matters not for this discussion that Silberman failed in his book to develop sufficiently these notions. After his stimulating first few chapters which talk in terms of a broadened definition of education, he then falls into the trap of considering only formal schooling. But, he attempted to do something in those initial chapters which we should applaud--to get at the real purpose of education. He stated that the aim of education should be to make moral men. And while one may not be entirely satisfied with this, Silberman is one of the very few who address themselves thoughtfully to the question of what education ought to achieve.

In sum, the past decade has witnessed a general disillusionment with regard to formal institutions of education. To some, the school has failed in its liberal promise to provide greater socio-economic equality; to others, it represents a potent instrument of oppression. Complementing these themes has been the idea that education should be treated as something much broader than mere schooling. On the whole, however, this critical spirit, while perhaps effective in analyzing the present educational system, lacks a reasoned sense of what ought to be. It criticizes without thinking through the consequences of that criticism. Nowhere, perhaps, is this lack of tough-minded thinking more apparent than in discussions of nonformal education, the educational fad which is a child of this current debate.

For the uninitiated let me describe briefly what seems to be the popular understanding and appeal of nonformal education. The first decade or so of independence for African nations witnessed an unprecedented growth in schools and universities. The amount of money and other resources devoted to educational expansion in Africa staggers the imagination. While figures vary from country to country, one fairly recent estimate is that African governments annually spend from 2% to 6.8% of their gross national products, and between 7% to 30% of their public expenditures, on education (M'Bow 1974). The largest single employer of trained manpower in African nations is the educational system. Yet, despite this expenditure of money and effort, the educational system seems never to be sated: if 50% of the children of primary school age are in school, so the cry is for 100%. As more and more primary school children complete their educations, there is pressure for more secondary schools; and the more secondary school graduates, for more university places. The costs are staggering. The acquiescence to such demands would compel African countries to devote their entire national budgets to education--and even then it is doubtful that the job would be done. Faced on the one hand with immense political pressure and on the other with the prospect of bankruptcy, it is logical that African governments have lately tried to find some cheaper alternative to the formal system of schooling.

The problem is not money alone; there is also an absence of trained manpower. Even if the money were there, the trained manpower needed to run an expanded educational system is not, and probably will not be for a generation or more. There are some stop-gap solutions, most notably the employment of numerous foreign nationals to staff secondary institutions and universities. But, in employing such outside specialists, African nations invariable return to the problem of cost. In sum, there is a temptation in Africa to turn to alternative supplies of educating manpower--men and women not hitherto considered as "teachers."

Beyond the problems of cost and personnel there looms a social issue: the "school-leaver." We all know the stereotype: a young man (strangely, never a young woman) who has received some modicum of formal education on the primary level and has, for one reason or another, left school. That little bit of education was a dangerous thing; it has alienated him from his traditional culture and given him the mistaken idea that he can "make it" in the modern sector. He has left the farm and gone off to the city where, unfortunately, he becomes unemployed, a problem for the nation and its economy. What is needed, argue the planners and politicians, is a different sort of education, one which will train young men in useful skills which they can practice within their traditional, rural environment.

Mix these problems well enough, and what you get is an alternative; nonformal education. Though there are some definitional issues which need to be resolved, the popular understanding of nonformal education is that it is education which takes place outside the formal school curriculum (it is sometimes referred to as "out-of-school" education). It is less an "education" in the liberal sense and more a training in employable skills. It can take place almost anywhere: in the home, where a father may teach his son farming skills; in a garage where a

mechanic may show an apprentice the intricacies of motor repair;
in a dry-cleaning establishment, where the owner consciously or
unconsciously may instruct his assistants in cleaning methods.
Note all the problems which nonformal education solves: it takes
the financial burden off the state to supply a useful education;
it makes use of a vast cadre of readily available "teachers;" it
gives young people a skill which they can use to get themselves
off the welfare dole and onto the tax roll. What could be
simpler?

But before we get too far, perhaps we ought to define our
terms more precisely. Defining "education" within the context of
this discussion is comparatively easy as the nonformal planners
appear to embrace the kind of expansive definition which Cremin
and others espouse. Defining what "nonformal" means is more
problematic. Recourse to the dictionary fails us as the word
"nonformal" is a neologism yet to be certified by Webster as a
real word. We can deduce, however, that the term can be taken to
mean "not formal." "Formal" is an adjective derived from the
word "form," which in turn means "a definite shape...a particular
structural condition of mode...the meter or style of arranging
and coordinating parts for a pleasing, effective result..." We
may infer, then, that something that is not formal does not have
"a definite shape..." and so forth. Thus, "nonformal education"
is a type of education which does not have an arrangement or
coordination of parts "for a pleasing or effective result..."
This is a play on words, perhaps, but it is not quite as amusing
when we consider the attention, time and resources devoted to a
concept whose very definition flies in the face of reason.
Consider, for example, the illogic of using the word "nonformal"
within the context of Cremin's definition of education. Some
might argue that a precise definition of terms is not necessary.
I remember asking one proponent of nonformal education what the
concept meant. We engaged in a not terribly clear dialogue for
a while until he finally said in exasperation, "I don't care

what we call it as long as we know what we mean!"

But are we in agreement on what we mean? Let us start with one extreme definition: all things which educate are non-formal. This would include cooperative agricultural schemes and on-the-job training in factories which are generally included in the nonformal category, but it would also include formal schooling and toilet-training which are generally not. We do not include formal education because it is precisely the thing which the nonformal approach seeks to challenge; toilet training is not included because, perhaps, it is too personal a form of learning. Let us narrow our definition: nonformal education includes all education which can and ought to be determined by public policy, with the exception of formal schooling. This definition is akin to that of David Abernathy (1972) who once suggested that nonformal education might be defined as that education disseminated by government in every ministry save that of education. Thus, it would include the education produced by television and radio, army training, and government programs for the dissemination of birth control information; and it would exclude what parents tell their children about the evils of smoking and what lovers learn in the back seats of cars. Once again, however, we have a problem as many nonformalites (put that in your dictionary) argue that much nonformal education can and ought to be conducted by agencies over which governments have little or no control. Consider, for example, an automobile repair shop where kids hang out and incidentally learn how to take apart an engine. Perhaps one criterion for nonformality might be that it prepares people for more productive employment, but then there are literacy programs which do not have that as a primary aim, but are still classified as nonformal. Let us make another stab at a definition: nonformal education consists of that education which ought to be determined by public policy, with the exception of formal schooling, together with some kinds of education which shall be determined to be nonformal.

Though we now have a working definition, we are not yet
ready to go on, for the nonformalites would add that it is not
sufficient to define and identify what already exists as nonformal
education; rather one should also suggest how nonformal education
can be made more effective. And how does one make it more effective? By formalizing it, of course! Let us review for a second.
The aim of nonformal programs is to get off the expensive, cre-
dential standard of the formal school system. With Illich, they
would like to break the monopoly of that system and thus free
education from the mere pursuit of parchment. They do realize,
however, that the utility and status of nonformal education needs
attention. Consequently, a primary aim of nonformal planners has
been to regularize, standardize and inspect the training given by
agencies of nonformal education. We even hear some nonformalites
saying that licenses and credentials validated by governments
should be issued to those who successfully complete training in,
say, automobile repair or dry cleaning. Thus, despite their
emotional attachment to the idea of de-schooling, one might argue
that the proponents of nonformal education are, in fact, school
men of the first order. Perhaps this is to be expected, as the
nonformalites are for the most part accustomed to societies of
the school.

There are societies, of course, which do approach education in a manner truer to the spirit of nonformal education. I
refer here to such nations as Cuba and the People's Republic of
China. These countries are unique in the comprehensive manner
in which they have organized their total societies in order to
achieve certain socialist and egalitarian goals. Because of the
striking success of nonformal education in Cuba and China, one is
reminded of an educational truism that the educational system of
a particular nation is a reflection of its overall social-economic
system. Hence, if one were serious about nonformalizing education, one must also champion the cause of a radical social transformation.

As the popular accounts have it, China and Cuba are each intent upon reducing individual competitiveness in their societies. They seek to break down existing class differences and to suppress the growth of "new classes." They stress the importance of cooperative rather than individual effort. They rely upon moral as opposed to material incentives. They seek to eliminate alienation in its many forms. Most important, they have a vision of man becoming more perfect than he is now. Whether the goal is to make "the new Cuban man" or "the new Socialist man," the practices of these two societies are predicated upon the assumption that man is plastic. He can be improved provided his socioeconomic circumstance undergoes the necessary transformation.

Thus, education plays a key role in these revolutionary societies: if man is to be remade, he must be reeducated. The educational programs of these societies are, in many respects, unlike those of other nations. Literacy is stressed in order to make men aware of their situation and knowledgeable with regard to social solutions. Work-study programs are yet another feature of these societies as they are mechanisms which train people in the dimensions of praxis and which combat the perpetuation or growth of social classes. Mao and Casto consciously attempt to make their entire societies a single educational system teaching the same national curriculum. Thus, they make use of numerous agencies for educational purposes, agencies ranging from the local political cell to the national press and media, from formal adult educational programs to informal contacts in factories and communes. Above all, the leaders of these revolutionary societies seek to educate by their own personal example, whether it be cutting sugar cane or swimming a river.

To be sure, there are institutions of formal education in those societies, but they are organized along lines different from those in most other countries. Credentials are downgraded in importance; they are no longer keys to the acquisition of power and material wealth. One gains admission to institutions

of higher education not so much on the basis of scholarly compe-
tence as demonstrated adherence to the egalitarian ideals of the
society. Teachers and scholars are no longer elites; rather they
are subjected to social criticism, made to descend from their
lofty status, and enjoined not only to serve but to become one
with the people. Indeed, each person is seen as a potential
teacher as well as student. A more Illichian world is hard to
imagine.

Admittedly, my description here glosses over many of the
problems of such revolutionary societies. It remains to be seen
whether one can indeed be induced to be at once "red" and
"expert"; it is still undecided whether human nature is as plastic as the revolutionary vanguard believes. Above all, it is not
yet clear whether the price in terms of lost individual liberty
is worth the prize in creating such extremely egalitarian, col-
lectivist societies. My intent here is not to debate these
highly contentious issues; rather, it is merely to assert that if
one wishes seriously to argue the case for nonformal education,
then he must be willing to urge a radical transformation of society. The two are inextricably bound up with each other. Social
innovations are not supermarket offerings: you cannot pick and
choose.

Consider as a sad case in point the literacy schemes of
Paolo Friere (1970). Friere was born in Recife, Brazil, which is
the center of one of the extreme situations of poverty in the
world. Though he extricated himself from that poverty and took a
Ph.D. in the philosophy of education, he never forgot his origins.
Beginning in the early 1960's, he developed and then taught in
northeast Brazil a highly effective literacy program whose philo-
sophical basis was a curious but appealing blend of John Dewey
and Karl Marx. Methodologically, he got away from the "Dick and
Jane reader" approach and instead taught reading from newspapers
highly critical of the government and the social order. Philo-
sophically, his aim was to use the written word as an instrument

to demonstrate that men could understand the forces which oppressed them and act effectively to change their social situations.
One might anticipate the rest of the story. When the military
took over in Brazil, Friere was first jailed and then exiled.
For a while he worked in Chile, but he is no longer welcome
there. Presently he is with the World Council of Churches in
Geneva and he frequently makes trips, as a consultant, to different nations in the developing world. He remains very much
respected in the world of literacy education because his techniques have demonstrated their effectiveness. But, it i·s doubtful that they will be employed on any widespread basis in the
developing world; you cannot borrow his techniques without also
having a willingness to live with their social consequences.

This seems like a simple enough lesson, but it has not
been learned by those who champion the cause of nonformal education. Instead, they seem willing to introduce nonformal education into the developing nations of Africa without thinking
through the possible serious consequences of their actions. And
what might these consequences be?

When one considers the tremendous expense governments
would encounter in formalizing, regularizing, and inspecting
nonformal education programs, the financial consequence might be
increased rather than lowered costs. One might conceivably argue
that introducing nonformal education would attract funds from
donor agencies which, of late, have been cutting back on assistance to the more traditional forms of education, since the donor
agencies seem to be beating the drums on nonformality. There is
the question, however, of the extent to which a foreign agency
should be heeded in deciding domestic educational policy.[1]

---

[1]The suspicious among us might argue that foreign donors
want nonformal education because they think it will be education
on the cheap, and thus it would relieve them of the more expensive
moral obligation to aid formal schooling. One might wonder also,

Second, it is doubtful that the recourse to nonformal education would reduce the need for trained educational manpower. There is a difference between a good mechanic and a good teacher of mechanics; and, in all probability, there are shortages of personnel who would make efficient, effective teachers in non-formal programs. Moreover, as the nonformal becomes formally organized, we might well expect that programs would have to be instituted to train the trainers. And this would take us back to square one: cost.

Perhaps more importantly, nonformal education would not solve as much as exacerbate the social ills which presently beset African nations. For the most part, the planners talk about non-formal education as an adjunct to, not a substitute for, the formal school system. The present popular demand is for every child to have the opportunity of formal schooling. Along come the nonformalites who sometimes argue that this is financially impracticable and socially undesirable, that for many or most of the young, education should take place outside the formal school setting. More usually, they argue that all young people should be given some form of formal education (the three R's) and then most of them should complete their educations on a nonformal basis. But, who is kidding whom? Who will be given the real rewards in society--the mechanic or the permanent secretary? And what will be the educational qualification needed to gain those rewards? Surely not a government license to strip engines and pump petrol. And who will be those people who go on in the formal school settings?

---

especially based upon the recent experiences of U.S.A.I.D., just how much donor money earmarked for nonformal education actually gets to the developing world. Much of it seems to have been spent in the United States where certain universities have been given handsome contracts to develop nonformal strategies.

In a private moment recently I had occasion to talk with
a dean of an African university faculty of education. Publicly,
he said, he felt obliged to mouth the shibboleth about how the
country needed nonformal education, how it needed mechanics and
farmers. But, not for his sons and daughters--they would go to
university. Of course, you do not need to be highly educated to
see the dean's point, and it is not lost on the people of Africa.
The establishment of nonformal education as a separate educational track for "other people" will be seen as precisely that.
If the reward systems of African societies are geared to those
of a competitive, corporate society which places a premium on
the possession of academic credentials, pressure will be maintained to increase formal school enrollments.

Which brings us back to the subject of the school. Much
has been written over the past few decades about how irrelevant
are the schools of Africa, a point which has been capitalized
upon by the nonformalites. As the rhetoric has it, formal
schooling alienates the student from his traditional way of life,
schools teach a curriculum foreign to the student's immediate
environment. Just how true this is today is debatable: there
is a point to be made that the fundamental role of the school in
Africa is precisely to alienate students from their traditional
way of life and to prepare them for life in a modern, corporate
state.

Which brings us back to the current debate over the role
of the school in the corporate order. The critics mentioned
earlier seem to agree that the schools do an effective job in
preparing students for that corporate order; what they take
issue with is the moral propriety of that order. A debate over
its propriety lies outside the scope of this paper--suffice
to say it is debatable. My point is that an interpretation one
might make of the critics' argument is that, if a nation is
committed to becoming a modern corporate society, then formal
schooling is vital to the achievement of that goal.

But we cannot say more than that. Much of the literature from Coleman to Illich argues that the schools really play little or no role in providing for individual or group mobility. Unfortunately, the data used to support this contention are derived almost exclusively from the developed world. There is evidence from the developing world, however, which just might modify this conclusion. For example, Philip Foster and Remi Clignet (1966) have argued that, at least for the present, schooling does provide for a considerable amount of social mobility in Africa. Perhaps after the passage of a generation and the establishment of an expanded class of the highly educated this may no longer be true; but this does not belie the fact that schooling in present-day Africa does allow social mobility true to the liberal, Jeffersonian sense. Moreover, one might also argue that, contrary to Coleman, the school could in the future be one of the key agencies which governments use to ensure a continuing equality of opportunity. To be sure, Coleman and his colleagues have argued that one's socio-economic status is determined more by one's familial origin and less by formal education. But, one may be loathe in a liberal, democratic society to have government interfere with private family matters no matter how laudable the ends. One can through public policy, though, affect what goes on in the schools as they are instruments of the state. As inefficient as the school is in achieving desirable social ends, it may be one of the best mechanisms available given the constraints imposed by a libertarian society.

In sum, as expensive and full of problems as formal schools are in Africa, there is much to be said in their favor. Africa has not yet reached the stage where it must grope for quixotic educational alternatives, like that child of our present disorienting confusion over education, nonformalites. Nonformal education is a malformed child which inherited in its genetic constitution the pathological soft-headedness of its intellectual forebearers.

REFERENCES

Abernathy, David. 1972. Unpublished ms. presented at the
African Studies Association Annual Meeting, Philadelphia.

Bailyn, Bernard. 1960. Education in the Forming of American
Society, (Chapel Hill, N.C.: University of North Carolina
Press).

Carany, Martin, (ed.). 1972. Schooling in a Corporate State:
The Political Economy of Education in America, (New York:
David McKay).

_____, 1974. Education as Cultural Imperialism. (New York:
David McKay).

Clignet, Remy, and Philip Foster. 1966. The Fortunate Few: A
Study of Secondary Schools and Students in the Ivory
Coast, (Evanston, Ill.: Northwestern University Press).

Coleman, James, et al. 1966. Equalities of Educational Opportunity, (Washington, D.C.: U.S. Government Printing
Office).

Cremin, Lawrence A. 1970. American Education: The Colonial
Experience, (New York: Harper and Row).

_____, 1971. "Curriculum Making in the United States,"
Teachers College Record, 73, 207-220.

Friere, Paolo. 1970. Pedagogy of the Oppressed, (New York:
Herder and Herder).

Grant, Gerald. 1973. "Sharing Social Policy: The Politics
of the Coleman Report," Teachers College Record, 75,
17-54.

Hodgson, Godfrey. 1973. "Do Schools Make a Difference?",
The Atlantic Monthly, March, 35-46.

Illick, Ivan. 1970. De-Schooling Society, (New York: Harper
and Row).

Jaeger, Werner. 1945. Paideia: The Ideals of Greek Culture,
(New York: Oxford University Press).

Karier, Clarence J., Paul Violas, and Joel Spring. 1973.
Roots of Crisis: American Education in the Twentieth
Century, (Chicago: Rand McNally).

Katz, Michael B. 1968. The Irony of Early School Reform: Educational Innovation in Mid-Nineteenth Century Massachusetts, (Cambridge, Mass.: Harvard University Press).

_____, 1971. Class, Bureaucracy and the Schools: The Illusion of Educational Change in America, (New York: Praeger).

M'Bow, Amadou Mahtar. 1974. "UNESCO at the Service of Education in Africa," Educafia, 1, 8-9.

Silberman, Charles. 1970. Crisis in the Classroom, (New York: Random House).

FOREIGN AND COMPARATIVE STUDIES PROGRAM
Maxwell School of Citizenship and Public Affairs
Syracuse University
119 College Place
Syracuse, N. Y. 13210

EASTERN AFRICAN STUDIES

II.	The Conflict Over What Is To Be Learned in Schools: A History of Curriculum Politics in Africa. 1971. 113 pp.	Stephen P. Heyneman	$ 4.50
IV.	Foreign Conflict Behavior and Domestic Disorder in Africa. 1971. 128 pp.	John N. Collins	4.50
V.	The Politics of Indifference: Portugal and Africa, A Case Study in American Foreign Policy. 1972. 41 pp.	John A. Marcum	2.50
VI.	A History of Agricultural Innovation and Development in Teso District, Uganda. 1972. 190 pp.	David J. Vail	5.50
VIII.	The Zimbabwe Controversy: A Case of Colonial Historiography. 1973. 142 pp.	David Chanaiwa	4.50
IX.	The Union of Tanganyika and Zanzibar: A Study in Political Integration. 1973. 114 pp.	Martin Bailey	4.50
X.	The Pattern of African Decolonization: A New Interpretation. 1973. 123 pp.	Warren Weinstein John J. Grotpeter	4.50
XI.	Islamization Among the Upper Pokomo. 1973. 166 pp.	Robert L. Bunger, Jr.	5.50
XII.	Protest Movements in Colonial East Africa: Aspects of Early African Response to European Rule. 1973. 96 pp. & xi.	Robert Strayer Edward I. Steinhart Robert Maxon	4.50
XIII.	Education for What? British Policy Versus Local Initiative. 1973. 100 pp. & viii.	Charles H. Lyons Kenneth J. King Richard D. Heyman John M. Trainor	4.50
XIV.	Ethnicity in Contemporary Africa. 1973. 59 pp. & vi.	Robert H. Bates	3.50
XV.	Class and Nationalism in South African Protest: The South African Communist Party and the "Native Republic," 1928-34. 1973. 67 pp.	Martin Legassick	3.50
XVI.	Africans in European Eyes: The Portrayal of Black Africans in Fourteenth and Fifteenth Century Europe. 1975. 98 pp.	Peter A. Mark	4.50
XVII.	Drought, Famine and Population Movements in Africa. 1975. 144 pp. & vi.	James L. Newman (Ed.)	4.50

XVIII. Technology for Socialist Development in Rural
        Tanzania. 1975. 64 pp.   David J. Vail   3.50

XIX.   Citizenship in Africa: The Role of Adult   Joel C. Millonzi   4.50
        Education in the Political Socialization
        of Tanganyika. 1975. 125 pp.

XX.   Health Care Financing in Tanzania.   Manuel Gottlieb   4.50
        1975. 110 pp.

XXI.   Three Aspects of Crisis in Colonial Kenya.   Bismark Myrick   4.50
        1975. 91 pp. & xxiii.   David L. Easterbrook
                                                          Jack Roelker

XXII.   Africa and International Crises.   Robert W. Brown   4.50
        1975. 106 pp.   Donald F. Heisel
                                                          Charles H. Lyons
                                                          Harvey Flad

XXIII. Political Conflict and Ethnic Strategies: A   Warren Weinstein   4.50
        Case Study of Burundi. 1975. 95 pp.   Robert Schrire

XXIV.   Becoming Ugandan: The Dynamics of Identity in   Marshall H. Segall   5.50
        a Modern African State. 1975. 200 pp.

XXV.   Eastern African Culture History   Joseph T. Gallagher   4.50
        1976. Approx. 100 pp.

EASTERN AFRICAN BIBLIOGRAPHIC SERIES

1. A Bibliography of Malawi. 1965. 161.   Edward F. Brown, Carol A. Fisher,   5.00
                                                John B. Webster

2. A Bibliography on Kenya. 1967. 461 pp.   John B. Webster, Shirin G. Kassam,   8.00
                                                Robert S. Peckham, Barbara A. Skapa

3. The Guide to the Kenya National Archives.   Robert G. Gregory, Robert Maxon,   13.00
    1969. 452 pp.   Leon Spencer

SPECIAL PUBLICATIONS

1. Basic Structure of Swahili. 1966. 151 pp.   James L. Brain   3.50

2. Modern Makonde Sculpture Exhibit Catalog.   Aidron Duckworth   4.00
    1968. 103 pp.

3. Shindano: Swahili Essays and Other   Johannes C. Mlela, Jean F. O'Barr,   2.50
    Stories. 1971. 55 pp.   Alice Grant, William O'Barr

4. Africana Microfilms at the E.S. Bird Library,   David L. Easterbrook   3.50
    Syracuse University: An Annotated Guide.   Kenneth P. Lohrentz
    1975. 72 pp.

OCCASIONAL BIBLIOGRAPHIES

3. A Bibliography on Anthropology and Sociology in Uganda. 1965. 60 pp. — Robert Peckham, Isis Ragheb, Aidan Southall, John B. Webster — 3.50

5. A Bibliography of Beuchuanaland. 1966. 58 pp. — Paulus Mohome, John B. Webster — 3.50

A Supplement to a Bibliography of Beuchuanaland. 1968. 32 pp. (formerly Occasional Bibliography #12; now combined with #5 and sold as one publication). — Paulus Mohome, John B. Webster, Catherine Todd

7. A Select, Preliminary Bibliography on Urbanism in Eastern Africa. 1967. 45 pp. — Barbara A. Skapa — 2.50

10. A Bibliography on Swaziland. 1968. 25 pp. — Paulus Mohome, John B. Webster — 2.50

13. A Supplement to a Bibliography of Malawi. 1969. 62 pp. — Paulus Mohome — 3.50

15. Education in Kenya Before Independence: An Annotated Bibliography. 1969. 196 pp. — L. A. Martin — 4.50

18. A Guide to the Coast Province Microfilm Collection, Kenya National Archives. Kenya Seyidie (Coast) Province, Correspondence & Reports, 1891-1962. 1971. 191 pp. — Harvey Soff — 4.50

19. Microfilms Related to Eastern Africa, Part I, (Kenya). A Guide to Recent Acquisitions of Syracuse University. 1971. 88 pp. — R. F. Morton, Harvey Soff — 3.50

20. A Supplement to a Select Bibliography of Soviet Publications on Africa. (1970-71). 1972. 27 pp. — Ladislav Venys — 2.50

21. Microfilms Related to Eastern Africa, Part II (Kenya, Asian and Miscellaneous). A Guide to Recent Acquisitions of Syracuse University. 1973. 142 pp. — David Leigh, R. F. Morton — 4.50

22. A Guide to the Nyanza Province Microfilm Collection, Kenya National Archives, Part I: Section 10B, Correspondence and Reports, 1925-1960. 1974. 137 pp. — Alan C. Solomon — 3.50

23. A Select Bibliography of Soviet Publications on Africa in General and Eastern Africa in Particular. (1972-73). 1974. 43 pp. (Supplement V) — Ladislav Venys — 2.50

24. A Guide to the Nyanza Province Microfilm Collection, Kenya National Archives, Part II: Section 10A - Correspondence and Reports, 1899-1942. 1974. 50 pp. — Alan C. Solomon — 3.50

25. A Guide to Nyanza Province Microfilm Collections, Kenya National Archives, Part III: Section 10, Daily Correspondence and Reports, 1930-1963, Vol. 1. 1975. 258 pp. & vi. — Alan C. Solomon, Kenneth P. Lohrentz — 5.50

26. A Guide to Nyanza Province Microfilm Collection,   Kenneth P. Lohrentz   5.50
    Kenya National Archives, Part III; Section 10,   Alan C. Solomon
    Daily Correspondence and Reports, 1930-1963,
    Vol. II.   1975.   254 pp. & vi.

27. Microfilms Relating to Eastern Africa, Part III   David L. Easterbrook   4.50
    (Kenya and Miscellaneous): A Guide to Recent   Alan C. Solomon
    Acquisitions of Syracuse University.   Thomas F. Taylor
    1975.   112 pp. & iv.

OCCASIONAL PAPERS

16. The Political Development of Rwanda and Burundi.   John B. Webster   4.50
    1966.   121 pp. (Biblio.)

21. Capital Expenditure and Transitional Planning   Gary Gappert   3.50
    in Zambia.   1966.   53 pp.

33. A Social Science Vocabulary of Swahili.   James L. Brain   2.50
    1968.   43 pp.

51. Basic Structure of Swahili, Part II.   James L. Brain   2.50
    1969.   34 pp.

51. A Short Dictionary of Social Science Terms   James L. Brain   3.50
    for Swahili Speakers.   1969.   70 pp.

52. Environment Evaluation and Risk Adjustment   James L. Newman (Ed.)   4.50
    in Eastern Africa.   1969.   53 pp.

53. The Pokot of Western Kenya 1910-1963: The   L. D. Patterson   3.50
    Response of a Conservative People to Colonial
    Rule.   1969.   54 pp.

57. National Liberation and Culture (1970 Eduardo   Amilcar Cabral   3.00
    Mondlane Memorial Lecture).   1970.   15 pp.

Publications may be ordered from: Publications Desk
                                   Foreign and Comparative Studies Program
                                   119 College Place
                                   Syracuse, N. Y. 13210

5/30/76

Maxwell School

Founded in 1924, the Maxwell School for fifty years has been training and educating young men and women for public service and academic careers. From the beginning the School has included all the University's social science departments, and this combination of professional and academic programs has enriched the content of all Maxwell's offerings. In addition to the traditional social science disciplines (Anthropology, Economics, Geography, History, Political Science, and Sociology), the School provides interdisciplinary degree programs in International Relations, Public Administration, the Social Sciences, and Urban and Regional Planning. Of the School's approximately 700 candidates for advanced degrees about half are in the traditional social science departments and the other half in interdisciplinary programs.

Today, the Maxwell School has 130 faculty members and approximately 700 students enrolled in graduate degree programs. Each year approximately 225 students receive master's degrees and 70 students Ph.D. degrees. In the three-year period, fall 1970 through spring 1973, Maxwell faculty members authored 78 books, 53 monographs, 296 articles, and 133 conference papers.